Research Reports ESPRIT

Project 1561 · SACODY · Volume 1

Edited in cooperation with
the Commission of the European Communities

J.-L. Faillot (Ed.)

Vibration Control of Flexible Servo Mechanisms

Springer-Verlag

Berlin Heidelberg New York
London Paris Tokyo
Hong Kong Barcelona
Budapest

Volume Editor

J.-L. Faillot
c/o Bertin
BP3, F-78373 Plaisir Cedex, France

ESPRIT Project 1561 "A High Performance Flexible Manufacturing System (FMS) Robot with On-Line Dynamic Compensation/Structure Allégée à Commande Dynamique (SACODY)" belongs to the Subprogramme "Computer Integrated Manufacturing and Engineering" of ESPRIT, the European Specific Programme for Research and Development in Information Technology supported by the Commission of the European Communities.

Project 1561 aims to develop a high performance robot for use in Flexible Automated Assembly System (FAAS) environments. The major design aims are to reduce the rigidity of the mechanical structure, while improving speed of operation and static and dynamic control. Solutions to the critical problems are implemented within a comprehensive software package permitting joint computer aided design of mechanical components and control laws. Innovative sensors improve global positioning accuracy and ensure tracking and vibration control. A generic industrial robot was used to validate the developed methodology, integrate innovative technologies, and illustrate the applicability of the concepts for real FAAS applications.

CR Subject Classification (1991): J.7, J.2, J.6, I.6.3, D.2.10

ISBN-13: 978-3-540-56142-2 e-ISBN-13: 978-3-642-77899-5
DOI: 10.1007/978-3-642-77899-5

Publication No. EUR 14414 EN of the
Commission of the European Communities, Dissemination of Scientific and Technical Knowledge Unit, Directorate-General Information Technologies and Industries, and Telecommunications, Luxembourg

Typesetting: Camera-ready by authors
45/3140 – 543210 – Printed on acid-free paper

Foreword

This volume reports the work carried out between 1987 and 1991 in the framework of the ESPRIT CIME Project 1561 entitled *A High Performance FMS Robot with On-Line Dynamic Compensation* more often referred to by its French acronym *SACODY*, standing for *Structure Allégée à COmmande DYnamique*.

The volume is the outcome of a collaborative R & D project performed by a European team coordinated by Bertin & Cie (France) and involving AEG AG and KUKA Roboter GmbH (Germany), LMS International and K.U. Leuven (Belgium) as well as University College Dublin (Ireland).

On behalf of this consortium, we would like to acknowledge the support of the Commission of the European Communities, without which the research and development reported hereafter would not have been possible. We would especially like to thank Mrs. Patricia Mac Connaill, Head of the ESPRIT CIME Division, Dr. Rainer Zimmermann, SACODY Project Officer, and the project reviewers Dr. Motta and Profs. Parker, Coiffet and Trostmann for the continuous interest they have shown for the project all along its life as well as for their precious advices.

We would also like to thank the different project partners for the collaborative spirit they have shown throughout the project, which has been a key factor for the success of SACODY. In addition, we acknowledge Prof. Van Brussel and his dynamic team, Dr. Wörn, Dr. Van Der Auweraer, Dr. Wunderlich and Prof. Connolly. These few lines may, we hope, serve to express the pleasure we had in coordinating their contribution to SACODY and towards the successful results presented in this Research Report.

September 1993 — The Editor

Table of Contents

Foreword V

Part I. Introduction

1. Introduction to the SACODY Project 3
   J-L. Faillot

2. Theoretical Aspects of Lightweight Robot Control 8
   J. Swevers, D. Torfs, M. Adams, J. De Schutter, H. Van Brussel, H. Thielemans, P. Vanherck

Part II. A Toolbox for Lightweight Robot Design and Verification

3. On the Identification of the Dynamic Characteristics of Flexible Robots 41
   H. Van der Auweraer

4. Dynamic Simulation of Multibody Flexible Servomechanisms 76
   G. Gallay, J-L. Faillot

5. CAD/CAE based Methodology for Optimising Robot Mechanical Design 95
   B. Duignan, P. Connolly

6. Sensor Systems for Robot Evaluation 112
   H. Van Brussel, F. Demeester, J. Van Den Bossche, P. Vanherck

Part III. Industrial Application of the SACODY Robot Control

7. Controller Design for a High-Performance Servo Level 151
   H. Wunderlich, H. Walter

8. Design and Implementation of an Antivibration Robot Control Software 169
   J. Deniard, J-L. Faillot, J. Swevers, D. Torfs

9. Robot Performance Testing 188
   H. Wörn, H-R. Tradt, M. Gerung

Conclusion and Outlook 205

PART I

Introduction

Chapter 1

Introduction to the SACODY Project

Jean-Luc Faillot
Bertin et Cie

1. Introduction

This volume addresses the achievements of the SACODY project carried out within the framework of the first phase of ESPRIT the European strategic programme for research and development in information technology of the Commission of the European Communities.

The primary aim of ESPRIT is concerned with improving the competitiveness of the European Information Technologies (IT) industry by promoting advanced IT-based solutions for use by industry and services. The programme covers five main areas namely *Microelectronics, Information Processing Systems & Software, Advanced Business & Home Systems, Computer Integrated Manufacturing & Engineering* (CIME) and *Basic Research* and a cross area activity, *the Open Microprocessor Systems Initiative*.

The strategic objective of the CIME area is to demonstrate how the manufacturing and engineering industry may benefit from the application and integration of Information Technologies in products and processes. It especially undertakes to promote the development and the integration of available and emerging Information Technologies in automation and control systems to achieve new levels in quality, functionality, flexibility and availability of control systems for production processes and manufacturing units.

The ESPRIT CIM project 1561 SACODY, carried out between 1987 and 1991, has comprehensively studied the problems linked with the control of lightweight robots, and has succeeded in demonstrating how the implementation of Computer Aided Testing and Dynamic Modelling Techniques enables the improvement of the accuracy of industrial robots while increasing their operational speed. Starting from a background mainly addressing Large Structures developed for Space applications, it has succeeded in transferring and applying a spatial control concept into the field of Industrial Robotics.

This volume is the outcome of a research work led by the six participants AEG AG, BERTIN, KUKA Roboter GmbH, the Catholic University of Leuven, LMS International and University College Dublin.

It proposes a monograph in which all the aspects of the SACODY lightweight robot integrated design, i.e. identification, modelling, control design and sensor systems, are extensively described in dedicated papers.

2. From Large Space Structures to Lightweight Industrial Robots

The tremendous progress made in space transportation during the sixties has contributed to launching in the early seventies space programs for the development for communication, astronomy, surveillance and space exploration purposes of very large satellites and spacecraft that had to be deployed, assembled and constructed after being launched into space by rockets or shuttles. The emergence of such systems, generically referred to as Large Space Structures, has led researchers to address for the first time the problems arising from the accurate control of mechanical systems featuring a high degree of mechanical flexibility as a consequence of the combination of large size and of drastic mass requirements. Firstly, Large Space Structures exhibit many low frequency and poorly damped vibration modes which make their design irrelevant on the sole basis of their rigid body characteristics. They are moreover submitted to stringent requirements in terms of shape, orientation, vibration suppression and pointing accuracy. Last, their space behaviour soon appeared to be very difficult to predict via on-earth testing.

The considerable effort brought during the seventies on the study of Large Space Structures gave birth to the new concept of Active Control of Flexible Structures: The behaviour and the performance of the structure were now to be tailored by means of sensors, actuators and on-line computer control.

Implicitly, a major step had been made in the field of servomechanisms design: It had moved from a partitioned concept, mainly built around a perfect -and rigid- mechanical design, towards an extremely interdisciplinary topic, in which the most advanced techniques in the field of structural mechanics, computer modelling and control science permanently interfere. In the early eighties, the new mechanisms design concepts were extensively devised, and a considerable amount of papers was devoted to the development of modelling methods and control systems for satellite antennae, solar arrays and other deployable appendages.

Approximately at the same period, the first papers addressing the problem of controlling the vibrations of lightweight manipulator-like structures appeared. The basic needs were again arising from robotic space applications, that are traditionally submitted to drastic minimum mass requirements for large range systems. But in this particular field of manipulator applications, the work done in majority by american laboratories was soon understood to be of major interest for industrial robotics: Increasing the operational speed of robots while improving their accuracy had been for years the main concern of robot manufacturers. They had until then dealt with this objective by stiffening the structures of their robots. Nevertheless, due to the continuous increase of speed requirements, robot designers had already investigated in depth the possibilities offered by a pure mechanical design approach: The price to pay for cost effective industrial robots consisted of bulky and energy-consuming structures able to carry only 3 to 5% of their total mass; building ever stiffer structures could no longer be contemplated.

From its start in 1987, the ESPRIT CIM SACODY project, using the same approach as the one used for Large Space Structures, set out to develop a new robot control strategy, relying on a better knowledge of robot structural dynamics, yielded by improved modelling and modal analysis techniques.

Despite the availability of the preliminary studies addressing the control of flexible motorized arms, already mentionned, the problem of controlling lightweight robots was at the beginning of SACODY an open issue, only addressed at a theoretical and methodological level, without real applicability considerations.

The work actually carried out within the framework of SACODY has allowed the definition of a new industrially applicable approach of robot control, up to ensure simultaneously fast motion of the robot and adequate control of the vibrations of its structure.

3. A Case Study: The Robotized Spot Welding

One could have thought that controlling the vibrations of industrial robots was only a problem to be faced in a future when robots are really lightweight and numerical controllers powerful enough to support complex antivibration control algorithms.

The purpose of the SACODY industrial demonstrator was to give a denial to this idea, by providing an experimental implementation of the project's outcome on industrial hardware. To this aim, the last phase of SACODY focused on the case of robotized spot welding. The reasons for such a choice are twofold:

Spot welding is the most developed robotic application, especially in the automotive industry, where spot welding robots represent *90%* of the robots at work in car body assembly operations. Moreover, process requirements are minimum, which enables the direct translation of spot welding robot performance improvements into clear productivity increases.

In addition, due to the combination of their large range and of the heavy tools they have to carry, spot welding robots very often suffer from a lack of mechanical rigidity in their joints. As it will be shown in this volume, this flexibility affects the quality of robot positioning, that is accompanied by tool vibrations all the more important as high robot accelerations are sought.

SACODY ended in July 1991 with the successful implementation of the project outcomes for designing the antivibration control system of the industrial demonstrator. The resulting servo control system was integrated to a prototype of industrial robot controller especially developed in the project. Improvements of *20%* to *50%* of the total robot positioning delay along with complete suppression of vibrations were achieved.

SACODY has finally paved the way for an integrated robot system design, allowing a reduction of moving masses, the expensive structural rigidity being replaced by advanced computer control techniques.

This approach is expected to result in a near future to the emergence of a new generation of robots, optimised in terms of improved performance, lightweight structures and cost-effectiveness.

4. Outline

This monograph addresses the different components of the integrated robot system design, sometimes referred to as mechatronic robot design.

First, it states *the basic problems* to be solved before engineering lightweight robots:

- What are the characteristics of lightweight robots?
- What are the specifications for their control?
- What are the tools and methods needed for designing their control?
- What are the algorithms to be implemented for meeting the specifications of lightweight robots control?

Chapter 2, beyond answering these basic questions, illustrates the proposed methodology by describing its application to laboratory systems especially built during SACODY at the Catholic University of Leuven (K.U.Leuven).

Then, this volume addresses in its second part *the tools* developed during the project for satisfying the requirements of *lightweight robot control design*.

Chapter 3 first details the LMS approach for solving the problems associated with the *identification* of the flexibilities and nonlinearities encountered in lightweight robots. After surveying the relevant techniques, it describes their application to laboratory systems and industrial robots.

Chapter 4 then describes the BERTIN general computer code ADAMEUS, dedicated to *modelling* and *simulating* non rigid multibody servomechanisms. After having described the general software characteristics and architecture, emphasis is once again put on the application of Adameus for modelling the SACODY industrial demonstrator. It is shown how the combined use of the LMS identification techniques and of the ADAMEUS code allows the elaboration of high precision numerical model, to be used for designing a high performance control.

Anyway, dynamic modelling techniques implementing software codes like ADAMEUS, may be used at an early design stage, as one component of a concurrent design of lightweight robots. Chapter 5 describes such an *integrated CAD/CAE* approach devised at University College Dublin (UCD) for optimizing the industrial robots design strategy.

Chapter 6 is devoted to the description of the developments carried out at the K.U.Leuven on two *sensor systems* for robot evaluation purposes. The first of these two systems, the RODYM system based on a digitizing tablet, has been extensively used all along the project execution for evaluating the performance of the different robot controllers tested in the project.

The last phase of the SACODY project has been dedicated to the design and implementation of the antivibration control on the industrial demonstrator. The third part of this volume is devoted to this integration phase. As above mentionned, the robot used for this demonstrator was a KUKA IR 160 spot welding robot. Nevertheless, the main goal of the final demonstration was to have antivibration control algorithms running on a prototype of industrial controller.

Chapter 7 details the development by AEG of the *Robot Controller prototype* for SACODY. After having recalled the SACODY controller development objectives, concentrated on a high performance controller servo level, it describes the technical features of the proposed hardware on which the antivibration algorithms were successfully implemented by BERTIN. The paper ends with an indication of the current and future applications of this servo level hardware.

Chapter 8 focuses on the *implementation of the antivibration control* on the AEG/KUKA demonstrator performed in collaboration between K.U.Leuven and BERTIN.

The volume ends in Chapter 9 by presenting and analysing the results of the *performance tests* carried out by KUKA with the different controller versions ranging from conventional controllers to the SACODY demonstrator. It describes the criteria retained by KUKA for evaluating the dynamic performance, and then quantifies the improvements brought by the SACODY antivibration lightweight robot control system.

Chapter 2

Theoretical Aspects of Lightweight Robot Control

Jan Swevers, Dirk Torfs, Marleen Adams, Joris De Schutter,
Hendrik Van Brussel, Hans Thielemans, Paul Vanherck

K.U. Leuven

1. Introduction

1.1. Problem Statement

Today's industrial robot controllers only use sensors, (mainly encoders and tachometers), which are collocated with the actuators. To position the robot end effector, the desired end effector position and orientation are converted by inverse kinematic computations into joint angles. The joints are then driven to these angles, each joint using a proportional, integral, and derivative (PID) controller based on collocated encoder and tachometer feedback. The robot is presumed stiff enough such that the end effector is in the intended position and orientation. The two main limitations of collocated controllers are: **1)** the robot links and joints must be very stiff, and therefore very heavy, in order to obtain some degree of precision. This limits the robot to slow speeds and/or requires high levels of drive power. **2)** The inherent flexibility in the structure (link and/or joint flexibility) makes it impossible to achieve truly high precision. Moreover, due to the ever increasing specifications in terms of speed, acceleration, and terminal accuracy, these robot structures and controllers do not satisfy anymore. New control strategies are needed, and the weight and inertia of the moving parts have to be decreased. This decrease leads to lower structural resonance frequencies and larger link and joint deformations, and therefore urges the need for new control algorithms. They prevent or damp out undesired oscillations, resulting from excitation of structural dynamics. But even for the present-generation industrial robots there is need for such control algorithms. Tests on industrial robots have shown that traditional robot controllers excite the structural resonance frequencies, and that inaccuracies due to static and dynamic deformations can be significant [1, 2, 3].

To achieve greater terminal precision, controllers which control the robot end effector are needed. These controllers are referred to as *flexible robot controllers* because they take into account the flexibilities in the robot structure. Flexible robot controllers use noncollocated sensors to measure directly or indirectly the end effector position. The price to pay to achieve this better performance is: **1)** extra costs due to extra sensors, and **2)** the problem of achieving stability. Noncollocated systems are hard to control: a good model of the system dynamics and rather sophisticated control algorithms are essential to achieve stability and good dynamic behaviour [4, 5, 6, 7, 8, 9, 10]. Therefore, the main steps in the development of a flexible robot controller are: **1)** choice of sensors, **2)** modelling and identification of the dynamics, and **3)** choice of control algorithms and calculation of control parameters. The following paragraphs discuss these different steps in more detail.

1.2. The Requirements for Controlling Non Rigid Robots

1.2.1. Choice of Sensors

Accurate end effector control requires direct or indirect measurement of the end effector position. The sensor type depends on the origin of flexibility in the system and on the application and environment. The most straightforward approach is direct end effector position measurement, for example by means of a 3D laser tracking system (see section 3.4) or a camera vision system [4]. This direct approach is however only feasible in laboratory environments or space applications. In industrial environments and applications, direct end effector position measurements are not realistic. Strain gauges or real-time optical link deflection measurement systems, such as DIOMEDES which has been developed in this project (section 3.4) [11, 12], are the appropriate sensors to measure link flexibility. A reliable and accurate joint deformation measuring technique is to use an extra encoder attached to the link in addition to the encoder mounted on the motor shaft [1]. The difference between the motor encoder signal and the corresponding link encoder signal is then a measure of the deformation of the joint. The latter technique is important, because for a wide class of industrial robots, experimental results reveal that joint flexibility rather than link compliance is the dominating source contributing of robot flexibility [1, 3, 13, 14, 15]. This technique is also applied in this project to control the flexibility of the first three joints of a KUKA IR 160/60 industrial robot.

1.2.2. Modelling and Identification

Generally, flexible robots are distributed parameter systems, and therefore have an infinite number of structural resonance frequencies. For practical purposes, only a limited number of these modes are modelled. Joint flexibility is in most cases modelled as a lumped torsional spring, resulting in one flexible mode per flexible joint [1, 6, 9, 14, 16, 17, 18, 19]. The number of modes taken into account for flexible links depend on the distribution of the resonance frequencies. Most researches deal with "very" flexible links which have many modes within a very limited frequency range, [4, 5, 8, 20, 21, 22, 23, 24]. In those cases, several flexible modes are modelled. For "less" flexible links there is often a large separation between the first and the second resonance frequency [7, 12, 25], such that only the first resonance frequency is taken into account.

Most researches base the control design on an *analytic model* of the robot, which is derived using Lagrange's equations or by applying d'Alembert-Jourdain's principle. Often two different and complementary methods are used to identify the parameters of these models: experimental identification methods and measurement of physical parameters [4, 6, 9, 16, 17, 18, 20, 26]. The measurement of the physical parameters of the robot, such as the link lengths, masses and inertias, spring constants, motor Coulomb friction, etc...., is in most cases the starting point of the identification. The experimental identification methods are used to complement, verify, or tune these data. These experimental identification methods all reduce to solving an overdetermined set of linear equations in the least squares sense. Hammer excitation [20], band limited random excitation [9, 18], and stepped sine excitation [4, 17, 18] are some of the techniques used to collect the experimental data which appear in these sets of equations.

Experimental test setups have a relatively simple construction, are known in detail and easily disassembled, such that accurate physical parameters can be obtained from measurements. For industrial robots the situation is different: they have a more complex construction and their disassembly is not straightforward. Therefore for industrial robots, experimental identification is the main source of information. It is the only approach to obtain information on for example, damping, joint stiffness, and friction [18]. For flexible multi-link robots, this identification approach is the only possible approach, because these systems are nonlinear.

For one-link flexible robots, which are linear systems, many other pure experimental identification methods exist. These identification methods calculate the parameters of state space models or input output models (transfer functions) from measured system input output data. In the framework of this project, a new approach for the identification of a flexible one-link orbot has been developed. It is based on a least squares frequency domain identification method which fits discrete time transfer functions to stepped sine measured frequency response functions. A high model accuracy is obtained by: **1**) dividing the total model of the flexible robot into two submodels of lower order and by identifying these two submodels separately, and **2**) by taking into account a priori parameter knowledge, such that the identification is restricted to the unknown part of the systems.

1.2.3. Control Design

In industrial robot applications, continuous path control has become more important. Continuous path control requires accurate tracking of a specified continuous path. It is used for applications such as arc welding, grinding, painting, and pasting. Point-to-point control requires only positioning accuracy. It is used for applications such as parts handling and spot welding. This difference has a consequence on the controller. Feedback can accomplish good positioning accuracy and is sufficient for point-to-point applications. It is not sufficient to guarantee fast and accurate tracking, since high feedback gains do not eliminate tracking errors. Feedforward has to be included to this end.

Linear control techniques apply to flexible one-links robots. Discrete time control is preferred to continuous time control because discrete time controllers take into account the time delay associated with the implementation on a digital computer. Continuous time controllers require very high sampling rates for stability. State feedback control is preferred to input-output control because state feedback enables the use of several sensor signals for the control signal calculation, which improves the dynamic performance of the feedback controller [4].

Several approaches are available for the state feedback gain and feedforward calculation. The classical techniques for the state feedback gain calculation are the *linear quadratic regulator* (LQR) approach and the *pole placement* approach. Pole placement has mainly been used in this project since it allows a fine tuning of the closed-loop dynamics. Since LQR control design methods address the optimal regulation problem, a good step response or nicely damped closed-loop dynamics is not a goal of this design method. Exact feedforward calculation involves a simulation with the inverse model [29, 31]. Nonminimal phase systems, such as the experimental test set up, cause problems with inverse model simulation, because unstable zeros become unstable poles in the inverse model. Most researchers avoid these instabilities by considering only steady state conditions for the feedforward calculation, or by restricting to acceleration and/or velocity feedforward. It is however more accurate to modify the inverse model according to Tomizuka's substitution scheme [29] and generate the feedforward signal from the modified inverse model.

Within the framework of this project, a new feedforward algorithm for accurate tracking control of nonminimal phase systems has been developped [30]. It is based on Tomizuka's substitution scheme, but adds extra feedforward terms to compensate for the error of the Tomizuka's algorithm. Experimental results on a one-link flexible robot show a drastic improvement of the new method over the Tomizuka's method [30].

The ultimate high performance controller for a multi-link flexible robot is a nonlinear controller, which is based on a complete nonlinear model of the robot and which uses sensors to calculate the complete system state. In the framework of this project, a *Generalized Nonlinear Decoupling* control method for controlling flexible robots, which are characterized by link elasticity, has been developped. Although this method is developped for flexible link robots, it can also be applied to flexible joint robots. The method consists of nonlinear feedback which linearizes the rigid body dynamics, followed by linear feedback of the full system state (both rigid body and flexible state vairables). Unlike most nonlinear control methods for flexible multi-link robots [10,6], this method has been tested on an experimental test setup (a two-link flexible robot), and on an industrial robot.

1.3. Chapter Outline

This chapter describes the design of a high performance continuous path tracking controller for two robots: an *experimental flexible one-link robot* and an *experimental flexible two-link robot*. In each case, the sensor choice, the modelling of the dynamics, the choice of the control algorithm and the design of the control parameters are explained in detail.

The identification and control of an experimental flexible one-link robot (section 3) is a first step in the development of a controller for more complex configurations, such as an industrial robot. Linear strategies, applied to the single degree of freedom, must be extended for the more complex case. To avoid unexpected problems, the characteristics of the one-link robot are as close as possible to those of a real robot. These characteristics include nonlinear joint friction and low amplitude oscillations. Moreover, the end-effector is not measured directly, but is estimated from the angular motor position and strain gauge measurements. Although strain gauges can not be applied to industrial robots, the strain gauges fulfil the same role in the modelling and control of the flexible one-link robot as the difference between the motor and link encoder signals for the industrial robot: strain gauge signals are a measure of the deformation of the flexible beam in the same way as the difference between the link and corresponding motor encoder signals are a measure of the joint deformation.

For simplicity, only one flexible mode is considered and precautions are taken to eliminate the "spillover", i.e. the destabilizing influence of the unmodelled modes on the control.

The generalized nonlinear decoupling controller, introduced in section 4, is tested on a two-link robot, the first joint of which is flexible. For the implementation on the two-link robot, the nonlinear decoupling controller is modified in two ways: The nonlinear linearizing feedback is replaced by a nonlinear linearizing feedforward, which reduces considerably the number of on-line calculations, and the full state feedback is based on a discrete model, obtained from the linearization of the remaining nonlinearities. Discrete time feedback is preferred to continuous time feedback: discrete time feedback controllers allow much lower sampling frequencies than continuous time feedback controllers because they take into account the time delay resulting from implementation on a digital computer. The controller is based on an analytic model of the test set up. The model form is derived using Lagrange equations. The joint flexibility is modelled as a massless torsional spring. The unknown parameters are then identified experimentally. The identification procedure is divided into three steps. The first step identifies the motor inertias and friction coefficients. Nonlinear Coulomb friction is included in this identification step. The second step identifies the spring constant associated to the flexible joint. The third step identifies the remaining parts of the nonlinear model. This identification procedure results in a much more accurate model than the straighforward identification of all unknown parameters from one set of equations.

Section 2 describes the linear regulation and tracking control law which has been implemented on the flexible one-link robot. Section 3 describes the identification and control design of a flexible one-link robot. It is shown that modes accuracy is essential for accurate tracking. Therefore, a lot of attention is paid to identification. Section 4 describes the generalized nonlinear decoupling control algorithm for flexible multi-link robots. Section 5 describes the implementation of that algorithm on a flexible two-link robot. Again, much attention is paid to accurate identification.

2. Independent Regulation and Tracking Control Law for Linear Systems

Decoupling the tracking and regulation parts of a control law is necessary to achieve high performance continuous path control: a feedforward takes care of tracking and reduces the continuous path control to a regulation problem. A feedback then eliminates small tracking errors due to environmental disturbances, modelling errors, etc..

The control design is based on a discrete-time state-space representation of the system:

$$X[k+1] = AX[k] + Bu[k] \quad (1)$$

$$y[k] = CX[k] \quad (2)$$

The feedforward u_{ff} is the input theoretically necessary to obtain an exact correspondence between the system's output y and desired trajectory y_d. This means:

$$X_d[k+1] = AX_d[k] + Bu_{ff}[k] \qquad (3)$$
$$y_d[k] = CX_d[k] \qquad (4)$$

$X_d[k]$ are the state variables obtained when there is an exact correspondence between the output $y[k]$ and the desired trajectory $y_d[k]$. Subtracting equation (1) from (3) and equation (2) from (4) and introducing the state error, $E[k] = X_d[k] - X[k]$ and $e[k] = y_d[k] - y[k]$, yields:

$$E[k+1] = AE[k] + Bu_e[k] \qquad (5)$$
$$e[k] = CE[k] \qquad (6)$$

with:

$$u_e[k] = u_{ff}[k] - u[k] \qquad (7)$$

System (5), described by the same state space model as (1), gives the relation between $E[k]$, $e[k]$, and $u_e[k]$. $u_e[k]$ must be calculated such that $e[k]$ is as small as possible. This is achieved with a regulation controller:

$$u_e[k] = -KE[k] \qquad (8)$$
$$\text{or } u_e[k] = -K(X_d[k] - X[k]) \qquad (9)$$

Combining equations (7) and (9) yields the control law for the original problem:

$$u_e[k] = u_{ff}[k] + K(X_d[k] - X[k]) \qquad (10)$$

It consists of two separate parts:

1. A feedforward u_{ff} that takes care of the tracking and is independent on the feedback,
2. a feedback of the difference between the desired and measured states that ensures the disturbances rejection by means of an appropriate set of feedback gains K.

From equations (3) and (4) it is clear that an exact feedforward calculation involves a simulation with the inverse model. Nonminimum phase systems, such as a flexible one-link robot, cause problems with inverse model simulation, because unstable zeros become unstable poles in the inverse model. Most researches avoid these instabilities by considering only steady state conditions for the feedforward calculation, or by restricting to acceleration and/or velocity feedforward [32], [33]. It is however more accurate to modify the inverse model and generate the feedforward signal from the modified inverse model [29]. The modification consists of replacing every unstable pole in the inverse model at $z = a$ with $|a| > 1$, by a stable zero at $z = a^{-1}$ and a pole at $z = 0$.

$$\frac{1}{z-a} \rightarrow \frac{1-az}{z(1-a)^2} \qquad (11)$$

This substitution does not introduce a phase error, but causes a gain error which increases for larger frequencies. In many applications this is of minor importance since the desired trajectory mainly consists of low frequency components. If the unstable zero is close to $z = 1$, the gain error becomes important. An addition of extra feedforward terms reduces the gain error for low frequencies and results in an acceptable tracking error (within the measurement accuracy of the non-collocated sensors). [29], [31], and [34] describe the feedforward calculation algorithm in detail.

The desired state trajectories $X_d[k]$ result from a simulation with the model using the feedforward signal as input (3).

Several criteria can be handled to calculate the state feedback gains. Pole placement is often the most appropriate approach since it allows a fine tuning of the closed-loop dynamics.

3. An Accurate Tracking Control for a Flexible One-Link Robot

This section describes the design of a state feedback controller for accurate continuous path following of an experimental flexible one-link robot. It consists of **1)** feedback of the difference between desired and measured states, **2)** feedforward, and **3)** integral control. The controller does not use a direct end point position measurement, but estimates this position from the masured motor angle and strain gauge signals. The feedforward calculation aims at zero tracking errors, is based on the inverse model, and does not depend on the feedback calculation. Integral control eliminates positioning errors caused by nonlinear motor friction. The model, on which uses stepped sine measured frequency response functions of the system. The algorithm divides the total model into two submodels in series, and estimates their parameters separately.

Section 3.1 describes the test setup along with some dynamic characteristics. Section 3.2 describes the frequency domain parameter identification algorithm developed for the test setup. Section 3.3 describes the control design, and discusses the control results.

3.1. Description of the Flexible One-link Robot

Figure 1 gives a schematic representation of the test setup. It consists of a flexible beam with a payload, connected to a flexible torsional beam, which itself is connected to a direct drive brushless DC-motor (maximum motor torque *200 Nm*). The control input to the system is a voltage between - 10 Volt and + 10 Volt, which is linearly converted by the power supply of the motor into a current. A built-in encoder with a resolution of 102400 pulses per revolution measures the angular motor position. The motor position is expressed in *radians*. Strain gauges on the flexible beam near to the axis of rotation, and in the middle of the torsional beam measure the beam deflections. The strain gauge signals are expressed in *μm/m*. The end effector position is not measured directly, but can be determined from the strain gauge and encoder signals. The end effector position is expressed in *mm* arc length. The first resonance frequency is a free-free mode in the neighbourhood of *5 Hz*. The second resonance frequency, which is at *38 Hz*, and all higher resonance frequencies are neglected in the identification and control. An anti-resonance exists at approximately *3 Hz*, which corresponds to a low-damping clamped-free mode. This mode is excited when, in the neighbourhood of the destination point, the motor velocity suddenly drops to zero due to dry friction (the static friction torque is about *6 Nm*, i.e. 3 % of the maximum torque), or when external forces are applied to the beam.

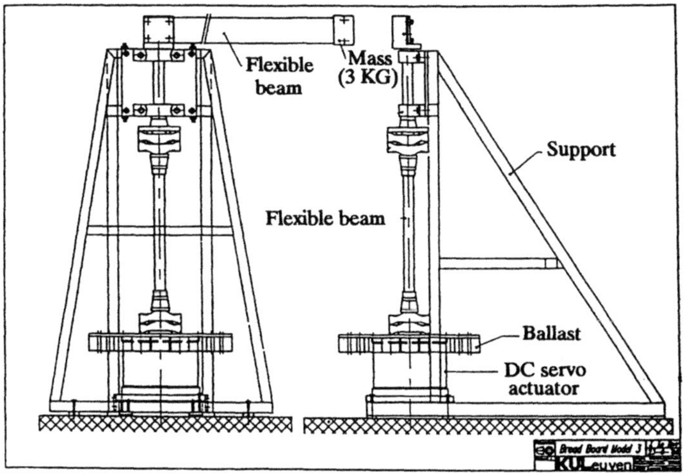

Figure 1: Schematic representation of the test setup

3.2. Frequency Domain Identification of a Flexible One-Link Robot

The identification of the total state space model is based on the identification of two discrete time transfer functions and on the determination of the relation between the end point position, the motor angle and the beam deflections. The division of the total model into two submodels and the separate identification of these two submodels based on two frequency response functions, result in a more accurate system model than the direct identification of the total model based on one frequency response function [32].

3.2.1. Mathematical Formulation of the Model

The form of the two discrete time transfer functions is determined from a physical analysis of the system based on [35]: the system is divided into rigid parts and linear elastic parts in which the mass and deformation properties are continuously distributed. The material of the elastic parts is assumed to be homogeneous and isotropic, and follows Hooke's law. Material damping is not considered, but linear motor friction is included. The analysis takes into account only the first resonance frequency (*5 Hz*). The second (*38 Hz*) and higher resonance frequencies are neglected in the modelling, identification, and control. This simplification makes the model only valid in a frequency range from *0* to *10 Hz*. The unmodelled modes may cause instability (spillover) during control. Section 3.2 discusses the precautions to overcome this problem.

The discrete time transfer functions that result from this analysis, and which are used to build the total state space model are:

$$H_1(z^{-1}) = \frac{b_{11}(z^{-1} - z^{-2})}{1 + a_{11}z^{-1} + a_{12}z^{-2} + a_{13}z^{-3}}$$
$$H_2(z^{-1}) = \frac{b_{20} + b_{21}z^{-1} + b_{22}z^{-2}}{1 - 2z^{-1} + z^{-2}} \qquad (12)$$

$H_1(z^{-1})$ relates the input u (proportional to the motor torque) to the strain gauge signal ϵ. $H_2(z^{-1})$ relates the strain gauge signal ϵ to the angular motor position α.

$H_1(z^{-1})$ contains a differentiation (one zero at $z = 1$), and $H_2(z^{-1})$ contains a double integration (two poles at $z = 1$).

3.2.2. Frequency Domain Parameter Estimation

The parameters of these two discrete-time transfer functions are identified with the frequency domain method presented in [36] and [37]. The method calculates the parameters of the models using stepped-sine-measured frequency response functions of the system. The estimation algorithm is based on the Singular Value Decomposition. The parameters of these transfer functions are estimated for a sampling frequency of *200 Hz*. The a priori information of the differentiator in $H_1(z^{-1})$ and the double integrator in $H_2(z^{-1})$ is explicitly taken into account during the parameter identification according to the technique discussed in [37]. It allows a more accurate estimation of the remaining parameters for less calculation time.

Figure 2 shows the amplitude and the phase of identified models $H_1(z^{-1})$ and $H_2(z^{-1})$ and of the measured frequency response functions. The resonance frequency is estimated in $H_1(z^{-1})$. Its value is *5.02 Hz* with a 3 % damping ratio. The anti-resonance frequency is estimated in $H_2(z^{-1})$. Its value is *2.99 Hz* with a 0.5 % damping ratio.

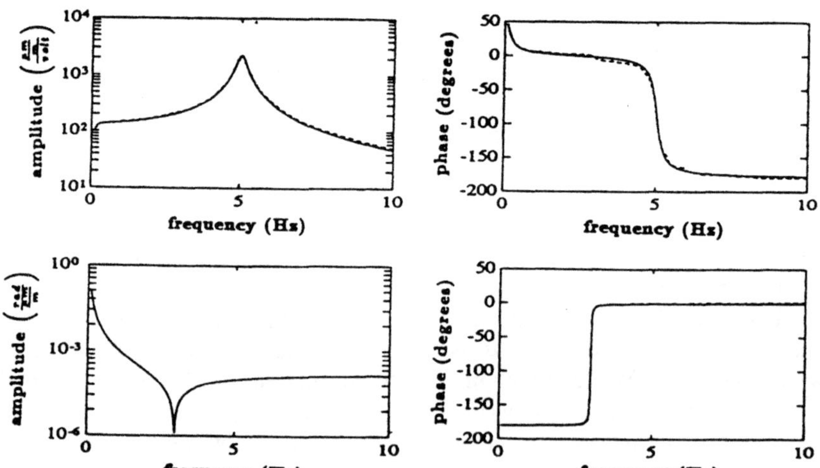

Figure 2
Comparison between the measured frequency response functions (dotted line) and the identified transfer functions (solid line)
top: $H_1(z^{-1})$
bottom: $H_2(z^{-1})$

3.2.3. Static Relation between End-Point Position, Strain-Gauge Signal and Motor Angle

The position $y[k]$ (in *mm* arc length) of the end effector is a linear combination of the strain gauge signal $\epsilon[k]$ and the motor angle $\alpha[k]$ if only the first mode is considered.

$$y[k] = a_{P1}\,\epsilon[k] + a_{P2}\,\alpha[k] \tag{13}$$

a_{P2} is equal to the length of the flexible beam in *mm*. a_{P1} has been determined experimentally. During the experiment the end point position is held constant, while the motor angle and the beam deflection are increased. The values of the strain gauge and encoder signals are measured for different beam deflections, and used to calculate a_{P1}.

3.2.4. Spillover

A flexible one-link robot has an infinite number of eigenfrequencies and anti-resonance frequencies. The identified model and the controller consider only the first of both. The unmodelled modes influence the reduced order controller and can even cause unstable behaviour [38]. Filtering the sensors outputs with digital low-pass filters reduce these spillover problems, since such filters remove higher order oscillations from the measured output signals. They extend the continuous-time physical system and the model with their discrete transfer functions. A second-order digital low-pass Butterworth filter with a *10 Hz* cut-off frequency has been used for this purpose. It has the following discrete transfer function:

$$H_q(z^{-1}) = \frac{2.08\,10^{-2} + 4.02\,10^{-2}z^{-1} + 2.08\,10^{-2}z^{-2}}{1 - 1.56z^{-1} + 6.41z^{-2}} \tag{14}$$

The discrete transfer function relating the input signal $u[k]$ to the filtered strain gauge signal $\epsilon_f[k]$ is a series combination of $H_1(z^{-1})$ and $H_q(z^{-1})$. The discrete transfer function relating the input signal $u[k]$ to the filtered encoder signal $\alpha_f[k]$ is a series combination of $H_3(z^{-1})$ and $H_q(z^{-1})$.

3.2.5. Total State-Space Model

The state-space model for the test setup is a series combination of three canonical state space models: the state space equivalent of the digital anti-spillover filter, the state space equivalent of $H_1(z^{-1})$ after elimination of the zero at $z = 1$, and the state space equivalent of $H_2(z^{-1})$.

A straightforward series combination of $H_1(z^{-1})$ and $H_2(z^{-1})$ results in a uncontrollable state space model (a common pole and zero at $z = 1$). The elimination of this common pole and zero makes the state space model controllable, but results in state variables that are equal to the integrated filtered strain gauge signal at the time instants k, $k - 1$, and $k - 2$. A small DC-offset on this signal, caused by the measurement equipment, results in an incorrect linear trajectory for these state variables, and a bad control performance. The elimination of the zero in $H_1(z^{-1})$ without elimination of the pole in $H_2(z^{-1})$ results in a controllable state space model, which has state variables that are equal to the filtered strain gauge signal at time instants k, $k - 1$, and $k - 2$, and filtered encoder signal at the time instants k and $k - 1$. This makes a state estimator unnecessary. The zero is cancelled by creating a new input signal: the differential input $\Delta u[k] = u[k] - u[k\text{-}1]$.

The total state-space model therefore relates the differential input $\Delta u[k]$ to the filtered end-effector position $y_f[k]$. It is presented in the following set of equations:

$$\begin{aligned} X[k+1] &= AX[k] + B\Delta u[k] \\ y_f[k] &= CX[k] \end{aligned} \qquad (15)$$

with:

$$X[k] = [x_1(k)\ \ x_2(k)\ x_3(k)\ \ x_4(k)\ \ x_5(k)\ \ x_6(k)\ \ x_7(k)]^T$$

$$[A] = \begin{bmatrix} 1.56 & -6.41 & 0 & 0 & 0 & 0 & 0 \\ 1 & 0 & 0 & 0 & 0 & 0 & 0 \\ 2.46\,10^{-1} & 2.48\,10^{-2} & 2.96 & -2.95 & 9.87\,10^{-1} & 0 & 0 \\ 0 & 0 & 1 & 0 & 0 & 0 & 0 \\ 0 & 0 & 0 & 1 & 0 & 0 & 0 \\ 3.77\,10^{-5} & 3.80\,10^{-6} & 1.49\,10^{-4} & -3.00\,10^{-4} & 1.52\,10^{-4} & 2 & -1 \\ 0 & 0 & 0 & 0 & 0 & 1 & 0 \end{bmatrix}$$

$$[B] = [\,1 \quad 0 \quad 6.90\,10^{-2} \quad 0 \quad 0 \quad 1.06\,10^{-5} \quad 0\,]^T$$

$$[C] = [\,0 \quad 0 \quad -1.32\,10^{-1} \quad 0 \quad 0 \quad 816 \quad 0\,]$$

The first submodel is of order two, and is the state space equivalent of the anti-spillover filter. It filters the differential input $\Delta u[k]$. The second submodel is of order three, and relates the filtered value of $\Delta u[k]$ to the filtered value of the strain gauge signal $\epsilon_f[k]$. The state variables of this model are $\epsilon_f[k]$, $\epsilon_f[k-1]$, and $\epsilon_f[k-2]$. The third submodel is of order two, and relates $\epsilon_f[k]$ to the filtered value of the motor angle $\alpha_f[k]$. The state variables of this model are $\alpha_f[k]$ and $\alpha_f[k-1]$. This model has an unstable zero at $z = 1.52$, which has to be substituted in the feedforward calculation according to equation (11).

3.3. Control Design and Tracking Control Results

The control design for the test setup is based on state space model (15). The input of this model is $\Delta u[k]$. This means that the signal from this controller must be integrated before it is applied to the system.

$$\begin{aligned} u[k] &= u(k\text{-}1) + \Delta u[k] \\ \text{with:} \qquad & \\ \Delta u[k] &= \Delta u_{ff}[k] - K(X[k] - X_d[k]) \end{aligned} \qquad (16)$$

Using this control law, the steady state position error measured at the end point amounts to *1.0 mm*, which is higher than the required error tolerance of *0.1 mm*. This error is due to the high static friction of the direct drive motor which makes the motor stop before the desired position is reached.

An integral term, based on a trapezoid integration rule, is added to the control input (16) to eliminate this steady state error:

$$u_i[k] = u_i[k\text{-}1] + k_i T_s\,(e[k] + e[k\text{-}1])/2 \qquad (17)$$

where k_i is the integral feedback gain and
T_s is the sampling period.

A tracking test has been executed to validate the controller performance. The reference trajectory is based on a tenth order polynomial. It gives intermediate positions between the initial and final position along a smooth path. The feedforward calculation method presented in this paper requires a smooth path: simulation with the inverse model of the system results in large peaks in the feedforward signal if the desired trajectory is not smooth. In addition it avoids unnecessary excitation of the eigenfrequencies.

The maximum end point acceleration and end point velocity during the tracking are $24\ ms^{-2}$ and $9\ ms^{-1}$. These values are near to the limitations of the motor. The total displacement of the end point is *5.13 m* arc length, which corresponds to one motor revolution. The final position should be reached after *1.555 seconds*.

A healthy approach to pole placement is to replace only the poles of the open loop system that result in a bad dynamic behaviour under tracking conditions. This avoids unnecessary large feedback gains. The two complex conjugated poles of the resonance frequency and the two poles at $z = 1$ are on or too close to the unit circle in the z-plane. They are not suited for a tracking control action and must be replaced. The system poles that originate from the Butterworth filter $H_q(z^{-1})$ result in an acceptable dynamic behaviour, and can be copied to the closed loop system. The choice of the remaining five pole positions is motivated by robustness considerations. Three poles are placed at $\omega_0 = 2\pi f\ rad\ s^{-1}$ and $\zeta = 1.0$ for good tracking dynamics, and two poles are placed at $\omega_0 = 2\pi f\ rad\ s^{-1}$ and $\zeta = 0.707$ for good regulator dynamics [39]. f is the frequency of the poles. These poles are dominant if f is smaller than the cut-off frequency of the Butterworth filter (*10 Hz*). A frequency f of *7 Hz* for the dominant poles of the controller resulted in the best performance for the tracking test.

The value of the integral feedback gain k_i is determined experimentally. Good results are obtained with $k_i = 1.6\ 10^{-2}$.

Figure 3 shows the control results. The horizontal dashed lines indicate the steady state error tolerance of *+/- 0.1 mm*. The vertical dashed line indicates the time instant at which the desired final position should be reached (*1.555 sec*). The maximum tracking error is less than *4 mm*. The amplitude of the small oscillations at the frequency of the anti-resonance frequency is less than *0.1 mm*. This corresponds to the resolution of the strain gauge signals. There is no overshoot, and the steady state error is within the tolerance of *+/-0.1 mm*.

An *0.085 kg* increase of the *3 kg* payload doubles the maximum tracking error and introduces a *0.3 mm* overshoot, but does not increase the small oscillations at the destination point. This proves that the model and feedforward signal are very accurate, but also that the tracking accuracy is very sensitive to modelling errors. The disturbance rejection ability, function of the feedback gains, is less sensitive to modelling errors.

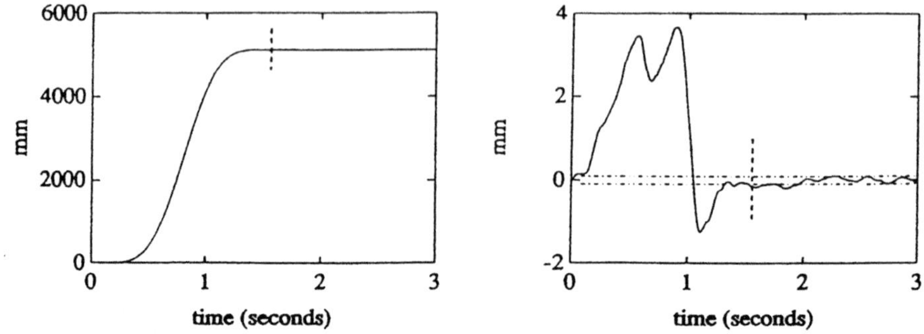

Figure 3: Control results.
Left: desired end-effector trajectory.
Right: tracking error.

3.4. Conclusion

The control approach presented in section 2 is tested on a flexible one-link robot. The controller for the test setup is based on an experimentally identified discrete-time state-space model. The tracking control tests prove that the proposed control design method results in accurate tracking, if the state space model on which the feedforward calculation is based, is accurate. From this it can be concluded that accurate feedforward calculation as well as accurate identification are necessary for fast and accurate tracking.

4. Control of Flexible Robots Using Generalized Nonlinear Decoupling

This section presents a new method for controlling flexible robots which feature link or joint elasticities. The method consists of:

- a nonlinear feedback to cancel out nonlinear terms in the rigid body dynamics;
- a linearization of the remaining nonlinearities, and linear feedback of all state variables (rigid and flexible coordinates). This ensures proper stability of the system about the desired state.
- Compensation of static and dynamic deflection improves the tracking of a reference trajectory. This compensation is achieved by calculation of the reference values of all state variables based on the model of the flexible robot.

It is shown how to select the linear feedback gains in order to obtain a closed-loop system with proper stability and dynamics.

4.1. Robot Model

The equations of motion of a manipulator with N rigid degrees of freedom θ_1, θ_2, ..., θ_N, to which are added P variables related to the flexibilities q_1, q_2,..., q_P, are given by:

$$\begin{bmatrix} M_{11}(\theta) & M_{12}(\theta) \\ M_{12}^T(\theta) & M_{22}(\theta) \end{bmatrix} \begin{bmatrix} \ddot{\theta} \\ \ddot{q} \end{bmatrix} = \begin{bmatrix} F_\theta(X) \\ F_q(X) \end{bmatrix} + \begin{bmatrix} T \\ 0 \end{bmatrix}, \tag{18}$$

where:

$$\theta = [\ \theta_1\ \theta_2 \ldots\ \theta_N\ ]^T$$
$$q = [\ q_1\ q_2 \ldots q_P\ ]^T$$
$$X = [\ \theta^T\ q^T\ \dot{\theta}^T\ \dot{q}^T\ ]^T$$

θ is a N-vector, q is a P-vector, and X is a $2(N + P)$-vector. T is the N-vector of torques applied at the joints.

$F_\theta(X)$ and $F_q(X)$ are nonlinear vector functions containing the effect of the Coriolis, centripetal and gravity forces, as well as the joints and links stiffness and damping effects. Since the mass matrix:

$$M(\theta) = \begin{bmatrix} M_{11}(\theta) & M_{12}(\theta) \\ M_{12}^T(\theta) & M_{22}(\theta) \end{bmatrix}$$

is symmetric definite positive, its inverse is also symmetric definite positive and is defined as:

$$W(\theta) = M(\theta)^{-1} = \begin{bmatrix} W_{11}(\theta) & W_{12}(\theta) \\ W_{12}^T(\theta) & W_{22}(\theta) \end{bmatrix}$$

The desired trajectory to be followed by the flexible manipulator is denoted by:

$$X_d = [\ \theta_d^T\ q_d^T\ \dot{\theta}_d^T\ \dot{q}_d^T\ ]^T$$

The tracking error $\Delta X = X - X_d$ is defined by:

$$\Delta X = [\ \Delta\theta^T\ \Delta q^T\ \Delta\dot{\theta}^T\ \Delta\dot{q}^T\ ]^T$$

4.2. Control Law Design

The most advanced controllers which exist for position control of rigid robots are based on computer torque, dynamic feedforward or nonlinear decoupling, which are all variants of the same idea. By compensation of the nonlinear terms in the rigid body dynamics, the system is reduced to a set of linear decoupled subsystems (1 subsystem per link). The same result can be obtained by applying the more general nonlinear decoupling method presented in [40]. The performance of controllers which are designed for rigid systems deteriorates very rapidly when the closed loop bandwith approaches the lowest structural resonance frequency (see simulation results in [27]).

Application of nonlinear decoupling to the complete system dynamics (rigid body plus flexible dynamics) does not bring a solution, since there are unsufficient independent control inputs in order to compensate for all the system's nonlinearities (there is only one actuator per joint, but there is an infinite number of degrees of freedom). The independent control inputs may be used to decouple and linearize the rigid body dynamics, but the flexible dynamics will then become uncontrollable.

The method presented in this section, called generalized nonlinear decoupling, also uses nonlinear decoupling (in an inner loop), but enables to control the full state of the system (in an outer loop).

The control law is split up into two feedback loops.

A first nonlinear inner loop realizes the feedback-linearization of the equations relative to the "rigid" degrees of freedom:

$$T = -F_\theta(X) - W_{11}^{-1}W_{12}(\theta)\, F_q(X) + W_{11}(\theta)\, T_1 \tag{19}$$

Equations (18) then become:

$$\ddot{\theta} = T_1$$
$$\ddot{q} = (W_{22} - W_{12}^T W_{11}^{-1} W_{12})(\theta)\, F_q(X) + W_{12}^T W_{11}^{-1}(\theta) T_1 \tag{20}$$

A second (outer) linear loop then ensures the stability of the system about the desired state X_d, by linear feedback of the full state X:

$$T_1 = \ddot{\theta}_d - K(X - X_d) \tag{21}$$

where $K = [K_\theta K_q K_{\dot{\theta}} K_{\dot{q}}]$ is a $N \times 2(N + P)$ feedback gain matrix.
Then, the closed-loop equations of the system become

$$\Delta\ddot{\theta} + K\Delta X = 0$$
$$\Delta\ddot{q} + W_{12}^T W_{11}^{-1} K\Delta X = (W_{22} - W_{12}^T W_{11}^{-1} W_{12})(\theta)\, F_q(X) + W_{12}^T W_{11}^{-1}(\theta)\ddot{\theta}_d - \ddot{q}_d \tag{22}$$

The complete feedback control law is obtained from equations (19) and (21) as:

$$T = -F_\theta(X) - W_{11}^{-1}W_{12}(\theta)\, F_q(X) + W_{11}(\theta)\, \ddot{\theta}_d - W_{11}^{-1}(\theta)K(X - X_d) \tag{23}$$

Figure 4 gives the corresponding block-diagram.

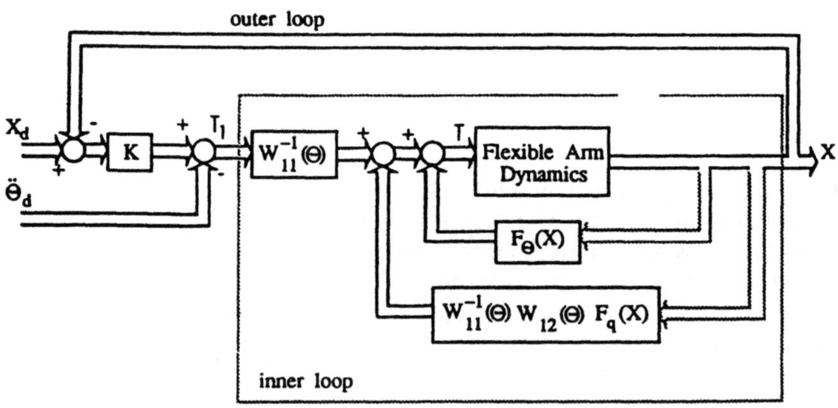

Figure 4: Block diagram of generalized nonlinear decoupling

4.3. Feedback Gains Computation

The feedback gain matrix must be designed from equations (22). In order to use linear techniques (pole placement, optimal control, ...) for feedback gains computation, the second equation of (22) is linearized to the first order about the desired trajectory $X_d(t)$:

$$W_{ij}(\theta) = W_{ij}(\theta_d),\ for\ i = 1, 2\ and\ j = 1, 2$$
$$F_q(X) = F_q(X_d) + \Delta F_q(X_d)\ \Delta X \qquad (24)$$

where $\Delta F_q(X_d)$ is a $P \times 2(N + P)$ matrix.

Denote

$$H(X_d) = (W_{22} - W_{12}^T W_{11}^{-1} W_{12})(\theta_d)\ \Delta F_q(X_d) \qquad (25)$$
$$H_0(X_d) = (W_{22} - W_{12}^T W_{11}^{-1} W_{12})(\theta_d)\ F_q(X_d) + W_{12}^T W_{11}^{-1} W_{12}(\theta_d)\ddot{\theta}_d - \ddot{q}_d \qquad (26)$$

Where $H(X_d) = [H_\theta(X_d)\ H_q(X_d)\ H_{\dot\theta}(X_d)\ H_{\dot q}(X_d)]$ is a $P \times 2(N + P)$ matrix, and $H_0(X_d)$ is a $P \times P$ matrix.

The linearized closed-loop equations can now be written, from equations (22), as

$$\Delta\dot{X} = \begin{bmatrix} 0 & 0 & I & 0 \\ 0 & 0 & 0 & I \\ -K_\theta & -K_q & -K_{\dot\theta} & -K_{\dot q} \\ A(4,1) & A(4,2) & A(4,3) & A(4,4) \end{bmatrix} \Delta X + \begin{bmatrix} 0 \\ 0 \\ 0 \\ H_0(X_d) \end{bmatrix}, \qquad (27)$$

where:

$$A(4, 1) = H_\theta(X_d) - W_{12}^T W_{11}^{-1} K_\theta$$
$$A(4, 2) = H_q(X_d) - W_{12}^T W_{11}^{-1} K_q$$
$$A(4, 3) = H_{\dot\theta}(X_d) - W_{12}^T W_{11}^{-1} K_{\dot\theta}$$
$$A(4, 4) = H_{\dot q}(X_d) - W_{12}^T W_{11}^{-1} K_{\dot q}$$

Inspection of equations (27) shows that the design of the feedback gains matrix:

$$K_d = [\, K_\theta\ K_q\ K_{\dot\theta}\ K_{\dot q}\,]^T$$

can be made by considering the following *linear time-varying tracking error system*:

$$\Delta\dot{X} = A\,(X_d)\,\Delta X + B\,(X_d)\,e + \delta(X_d) \qquad (28)$$

where:

$$A(X_d) = \begin{bmatrix} 0 & 0 & I & 0 \\ 0 & 0 & 0 & I \\ 0 & 0 & 0 & 0 \\ H_\theta(X_d) & H_q(X_d) & H_{\dot\theta}(X_d) & H_{\dot q}(X_d) \end{bmatrix}$$

$$B(X_d) = \begin{bmatrix} 0 \\ 0 \\ I \\ W_{12}^T W_{11}^{-1}(\theta_d) \end{bmatrix}$$

$$\delta(X_d) = \begin{bmatrix} 0 \\ 0 \\ 0 \\ H_0(X_d) \end{bmatrix}$$

$$\delta(X_d) = \begin{bmatrix} 0 \\ 0 \\ 0 \\ H_0(X_d) \end{bmatrix}$$

A linear time-varying state feedback:

$$e = -K \, \Delta X \tag{29}$$

applied to the system described by equations (28), will solve the control problem of the system described by equations (22).

4.4. Conclusion

A method is presented for controlling multilink flexible manipulators. It consists of nonlinear feedback to linearize and decouple the rigid body dynamics, followed by linear feedback of the full system state. It is shown how the feedback gain matrix has to be selected in order to obtain proper stability and closed loop dynamics.

5. Nonlinear Control of a Flexible Two-link Robot

This section reports on experimental experience with identification and control of a two link robot with a flexible first joint. The controller is a slightly modified version of the generalized nonlinear decoupling controller presented in section 4. The nonlinear feedback is replaced by a nonlinear feedforward. This means that the nonlinear decoupling terms are calculated based on the desired state variables instead of the actual values. The advantage is that the nonlinear decoupling torque can be calculated off-line, and that using the noise free desired values results in more stable behaviour of the controller.

The controller is based on an analytic model of the robot. The experimental identification of this model is described in section 5.1. The high nonlinear Coulomb friction in the system is taken into account in the identification. This nonlinear friction is also compensated during the control. Section 5.2 describes the control design. Section 5.3 describes the experimental results. The performance of the nonlinear controller is compared with that of a classical rigid controller. The tests show the superior behaviour of the nonlinear controller.

5.1. Experimental Setup

5.1.1. Description of the Test Setup and Mathematical Formulation of the Model

Figure 5 gives a schematic respresentation of the test setup. The real setup is shown in figure 6. It consists of a series combination of two beams driven by direct drive motors.

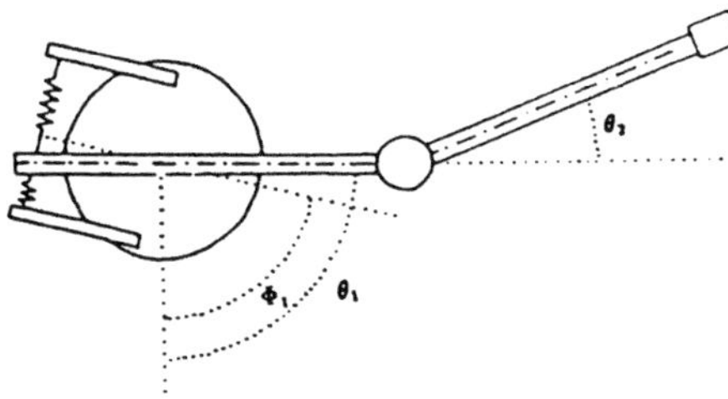

Figure 5: Schematic representation of the test setup

The motion takes place in the horizontal plane. The first beam is connected by springs to the motor. This simulates a flexible first joint. The first resonance frequency varies from *4.6 Hz* to *4.8 Hz* as a function of the position of the second link. The motor positions are measured by encoders, the deformation of the spring is measured by a LVDT (linear variable differential transformer). The control input to the system is a voltage between - *10 Volt* and + *10 Volt*, which is converted by the power supply of the motor into a current. The motor torque is proportional to the current.

An essential information for the control design is the knowledge of the total dynamic model.

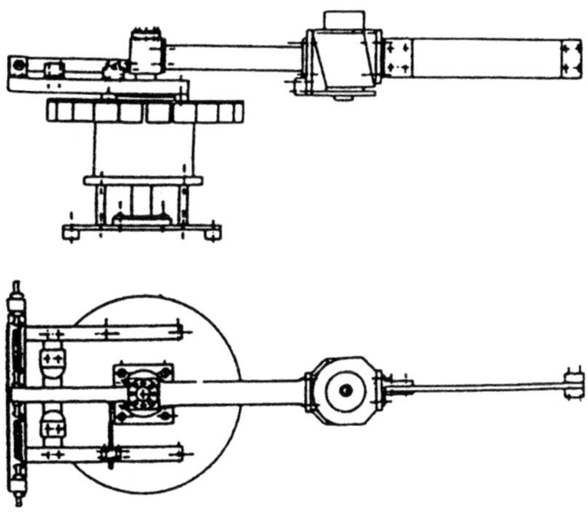

Figure 6: The real test setup

Applying Lagrange's equations yields the system's nonlinear equations of motion:

$$I_{m1}\ddot{\Phi}_1 + T_{f1} - K_s(\theta_1 - \Phi_1) = T_1 \tag{30}$$

$$(M_2 + M_c\, cos(\theta_2))\, \ddot{\theta}_1 + M_2\, \ddot{\theta}_2 + M_c\, \dot{\theta}_1^2 sin(\theta_2) + T_{f2} = T_2 \tag{31}$$

$$\begin{aligned}(M_1 + 2\, M_c\, cos(\theta_2))\, \ddot{\theta}_1 + (M_2 + M_c\, cos(\theta_2))\, \ddot{\theta}_2 \\ -2\, M_c\, \dot{\theta}_1 \dot{\theta}_2 sin(\theta_2) - M_c\, \dot{\theta}_2^2 sin(\theta_2) + K_s(\theta_1 - \Phi_1) = 0\end{aligned} \tag{32}$$

where

Φ_1	:	motor position 1
θ_1	:	beam position 1
$\theta_1 - \Phi_1 = q_1$	:	deformation of the spring
θ_2	:	beam position 2 = motor position 2
T_1	:	motor torque 1
T_2	:	motor torque 2
I_{m1}	:	rotor inertia 1
T_{f1}	:	motor friction torque 1
T_{f2}	:	motor friction torque 2
K_s	:	spring constant
M_2	=	$I_{m2} + I_{b2} + I_p + m_p l_2^2$: Total inertia 2
M_1	=	$I_{b1} + I_{b2} + I_{m2} + I_p + m_{b2} l_1^2 + m_{m2} l_1^2 + m_p l_1^2 + m_p l_2^2$
M_c	=	$l_1 l_2 (\frac{m_{b2}}{2} + m_p)$

Initially, it is assumed that the motor frictions consist only of linear viscous frictions:

$$T_{f1} = C_{m1} \dot{\Phi}_1$$

$$T_{f2} = C_{m2} \dot{\theta}_2$$

5.1.2. Identification

The unknown parameters in this model have to be identified. Because the dynamic model is nonlinear, the classical identification schemes based on a linear input-output model can not be used. Therefore the following procedure is applied: a known input torque is applied to the system and the corresponding positions and deformations of the spring are measured. Introducing the measured values and their first and second time-derivatives in the non linear equations gives rise to an overdetermined set of linear equations. For example, equation (30) becomes:

$$\begin{bmatrix} \ddot{\Phi}_1(i) & \dot{\Phi}_1(i) & -(\theta_1(i) - \Phi_1(i)) \end{bmatrix} \begin{bmatrix} I_{m1} \\ C_{m1} \\ K_s \end{bmatrix} = T_1(i) \qquad i = 1, \ldots, N$$

Where N is the number of measurement instants and $[I_{m1}\ C_{m1}\ K_s]^T$ the unknown parameter vector. This set is solved in a linear least squares sense to obtain the unknown parameters of the model.

In practice the identification process is divided into different steps.

Step 1: Identification of the motor inertia and friction coefficients

The measurements on the direct drive motors are performed on two separate setups. The dynamic behaviour is described by:

$$\begin{aligned} T_1 &= I_{m1}\ddot{\Phi}_1 + C_{m1}\dot{\Phi}_1 \quad \text{(motor 1)} \\ T_2 &= I_{m2}\ddot{\theta}_2 + C_{m2}\dot{\theta}_2 \quad \text{(motor 2)} \end{aligned}$$

It turns out that the Coulomb friction is not negligible. Therefore the total motor friction is modelled as:

$$T_{f1} = C_{m1}\dot{\Phi}_1 + F_{m1} sgn(\dot{\Phi}_1) \quad \text{(motor 1)} \tag{33}$$
$$T_{f2} = C_{m2}\dot{\theta}_2 + F_{m2} sgn(\dot{\theta}_2) \quad \text{(motor 2)} \tag{34}$$

where
F_{m1} : Coulomb friction of motor 1
C_{m1} : linear viscous friction of motor 1
F_{m2} : Coulomb friction of motor 2
C_{m2} : linear viscous friction of motor 2

The parameters C_{m1}, C_{m2}, F_{m1} and F_{m2} are identified with the following measurement procedure [17]: After bringing the motor to a certain velocity the torque is put to zero. While the velocity is decaying to zero the position is measured. The dynamic behaviour is described by:

$$-I_{m1}\,\ddot{\Phi}_1 = C_{m1}\dot{\Phi}_1 + F_{m1}\, sgn(\dot{\Phi}_1) \quad \text{(motor 1)}$$
$$-I_{m2}\,\ddot{\theta}_2 = C_{m2}\dot{\theta}_2 + F_{m2}\, sgn(\dot{\theta}_2) \quad \text{(motor 2)}$$

In these equations the identified rotor inertia are supposed to be exactly known. Solving this equation in a least squares sense gives the values of the friction coefficients. The measured and identified relationship between velocity and friction torque are compared in figure 7.

Step 2: Equation (30) is identified.

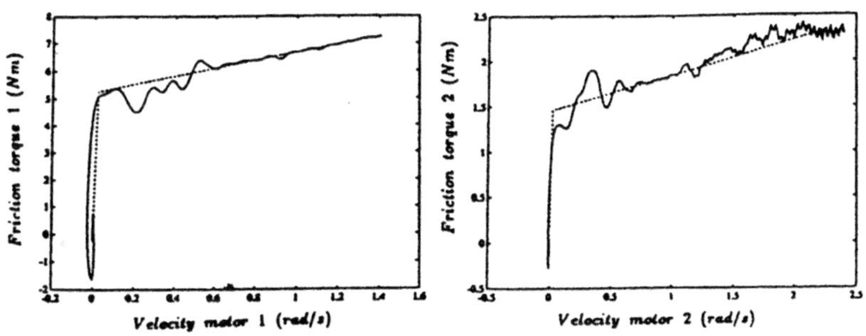

Figure 7: measured (solid line) and identified (dashed line) relationship between velocity and friction torque.

Step 3: Equation (31) and (32) are solved together. The motor friction, and the already identified spring constant are considered as known parameters: this means that the terms in these parameters are put in the right hand side of the equations.

This identification process is repeated for different noise measurements. Table 1 shows the coincidence between the identified and analytical values, together with the deviation on the identified parameters.

Analytical parameters		Identified parameters	deviation
I_{m1}	$1.6\ kgm^2$	$1.936\ kgm^2$	2 %
C_{m1}	–	$1.4\ kgm^2/s$	–
F_{m1}	–	$5.28\ Nm$	–
C_{m2}	–	$0.4\ kgm^2/s$	–
F_{m2}	–	$1.37\ Nm$	–
K_s	$1150\ Nm/rad$	$1223\ Nm/rad$	1.3 %
M_2	$0.915\ kgm^2$	$0.853\ kgm^2$	9 %
M_1	$7.17\ kgm^2$	$6.42\ kgm^2$	2.6 %
M_c	$1.085\ kgm^2$	$1.05\ kgm^2$	7 %

Table 1:
Comparison between the identified and analytical parameters of the model and deviation on the identified parameters for noise measurements

5.2. Control System Design

5.2.1. Control Model

For control design purposes the model (equations 30, 31 and 32) is put in the suitable form:

$$M(\theta_2)\begin{bmatrix}\ddot{\Phi}_1\\ \ddot{\theta}_2\\ \ddot{q}_1\end{bmatrix}+\begin{bmatrix}F_{\phi_1}\\ F_{\theta_2}\\ F_{q_1}\end{bmatrix}=\begin{bmatrix}T_1\\ T_2\\ 0\end{bmatrix} \tag{35}$$

with

$q_1 = \theta_1 - \Phi_1$,

$$\begin{aligned}M(\theta_2) &= \begin{bmatrix} I_{m1}+M_1+2\,M_c\,cos(\theta_2) & M_2+M_c\,cos(\theta_2) & M_1+2\,M_c\,cos(\theta_2)\\ M_2+M_{12}\,cos(\theta_2) & M_2 & M_2+M_c\,cos(\theta_2)\\ M_1+2\,M_c\,cos(\theta_2) & M_2+M_c\,cos(\theta_2) & M_1+2\,M_c\,cos(\theta_2)\end{bmatrix}\\ &= \begin{bmatrix} M_{11}\,(2\times 2) & M_{12}\,(2\times 1)\\ M_{12}^T\,(1\times 2) & M_{22}\,(1\times 1)\end{bmatrix}\end{aligned}$$

and

$$\begin{aligned}F_{\phi_1} &= C_{m1}\dot{\Phi}_1 - 2\,M_c\,\dot{\theta}_1\dot{\theta}_2 sin(\theta_2) - M_c\,\dot{\theta}_2^2 sin(\theta_2)\\ F_{\theta_2} &= C_{m2}\dot{\theta}_2 + M_c\,\dot{\theta}_1^2 sin(\theta_2)\\ F_{q_1} &= -2\,M_c\,\dot{\theta}_1\dot{\theta}_2 sin(\theta_2) - M_c\,\dot{\theta}_2^2 sin(\theta_2) + K_s q_1\end{aligned}$$

The state vector is denoted by:

$$x_d = [\Phi_{1d}\ \ \theta_{2d}\ q_{1d}\ \dot{\Phi}_{1d}\ \dot{\theta}_{2d}\ \dot{q}_{1d}]^T$$

The desired state vector is denoted by:

$$x = [\Phi_1\ \ \theta_2\ q_1\ \dot{\Phi}_1\ \dot{\theta}_2\ \dot{q}_1]^T$$

5.2.2. CONTROL LAW

The control law is split up into two stages.

First stage: a non linear feedforward which realizes linearization of the state variables containing the rigid part:

$$\begin{bmatrix} T_1 \\ T_2 \end{bmatrix} = \begin{bmatrix} F_{\Phi_1}(x_d) \\ F_{\theta_2}(x_d) \end{bmatrix} + W_{11}^{-1}(\theta_{2d})W_{12}(\theta_{2d})\ F_q(x_d) + W_{11}^{-1}(\theta_{2d})\begin{bmatrix} T_{fb1} \\ T_{fb2} \end{bmatrix} \qquad (36)$$

with

$$W = M^{-1} = \begin{bmatrix} W_{11}\ (2\times 2) & W_{12}\ (2\times 1) \\ W_{12}^T\ (1\times 2) & W_{22}\ (1\times 1) \end{bmatrix}$$

If the real state variables are approximately equal to the desired values equation (35) become:

$$\ddot{\Phi}_1 = T_{fb1}$$
$$\ddot{\theta}_2 = T_{fb2}$$
$$\ddot{q}_1 = (W_{22} - W_{12}^T W_{11}^{-1} W_{12})(\theta_{2d})\ F_{q1}(x) + W_{12}^T W_{11}^{-1}(\theta_{2d})\begin{bmatrix} T_{fb1} \\ T_{fb2} \end{bmatrix}$$

Second stage: a linear feedback of the full state x, which ensures proper stability of the system about desired state x_d:

$$\begin{bmatrix} T_{fb1} \\ T_{fb2} \end{bmatrix} = \begin{bmatrix} \ddot{\Phi}_{1d} \\ \ddot{\theta}_{2d} \end{bmatrix} - K(x - x_d) \qquad (37)$$

Where K is a 2 x 6 feedback gain matrix.

Figure 8 gives the corresponding block-diagram. In order to use linear techniques (pole placement, optimal control) for feedback gains computation, the equation (32) is linearized to the first order about the desired trajectory x_d. The feedback gain computation is explained in section 4. To limit the computation time, the feedback gains are chosen constant: they are computed with respect to the final desired position.

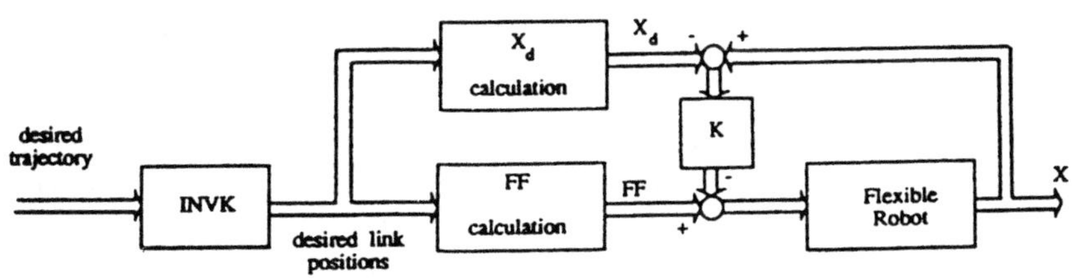

Figure 8: Block diagram of the nonlinear controller

5.2.3. TRAJECTORY GENERATION

In order to limit the excitation of the structure, a smooth trajectory along a straight line in the x-y plane is generated based on a ninth order polynomial. This straight line motion of the tip is transformed to the angular position, velocity and acceleration of each link, θ_{1d}, θ_{2d}, $\dot{\theta}_{1d}$, $\dot{\theta}_{2d}$, $\ddot{\theta}_{1d}$, $\ddot{\theta}_{2d}$, by using of the inverse kinematic equations of a rigid two-link robot [45].

In the control the tracking error $\Delta x = x - x_d$ is fed back. Therefore the reference values of the state variables

$$x_d = [\Phi_{1d} \;\; \theta_{2d} \; q_{1d} \; \dot{\Phi}_{1d} \; \dot{\theta}_{2d} \; \dot{q}_{1d}]^T$$

are needed. Based on the model (equation 32), the reference values for the deflection at the first joint, q_{1d} and the joint angle Φ_{1d} can be calculated:

$$\begin{aligned} q_{1d} = & \;(-(M_1 + 2\, M_c\, cos(\theta_{2d}))\, \ddot{\theta}_1 - (M_2 + M_c\, cos(\theta_{2d}))\, \ddot{\theta}_2 \\ & +2\, M_c\, \dot{\theta}_{1d}\dot{\theta}_{2d} sin(\theta_{2d}) + M_c\, \dot{\theta}_{2d}^2 sin(\theta_{2d})\;)\; /\; K_s \\ \Phi_{1d} = & \;\theta_{1d} - q_{1d} \end{aligned}$$

Differentiation gives the derivatives $\dot{\Phi}_{1d}$ and $\dot{q}_{1d}$.

5.2.4. Nonlinear State Estimator

The nonlinear controller is a continuous-time controller and it is assumed that all the states are available for feedback. However the velocities $\dot{\Phi}_1$, $\dot{\theta}_2$ and $\dot{q}_1$ can not be measured directly. Instead of using a differentiating algorithm, which can yield stability problems, a discrete nonlinear state estimator is used. It is a current estimator consisting of two stages: a measurement update:

$$\hat{x}_k = \bar{x}_k + L(y_k - \bar{y}_k)$$

and a time update:

$$\bar{x}_{k+1} = \int_{t_k}^{t_{k+1}} f(\hat{x}_k, T_{1k}, T_{2k})dt$$

with

$$f(\hat{x}_k, T_{1k}, T_{2k}) = \begin{bmatrix} \hat{x}_4 \\ \hat{x}_5 \\ \hat{x}_6 \\ -M^{-1}(\hat{x}_k) \begin{bmatrix} F_{\Phi_1}(\hat{x}_k) \\ F_{\theta_2}(\hat{x}_k) \\ F_{q_1}(\hat{x}_k) \end{bmatrix} + M^{-1}(\hat{x}_k) \begin{bmatrix} T_{1k} \\ T_{2k} \\ 0 \end{bmatrix} \end{bmatrix}$$

The unstable differential operation is thus avoided and replaced by a more stable integration. The estimator structure is similar to that of a linear current estimator. The only difference is that the time update of the states is provided by integration of the full non linear model. The estimator gain matrix, L, is calculated based on the linearized and discretized model. Only one constant feedback gain matrix is chosen. This matrix is designed with respect to the final desired position.

The controller stability is provided by:

- a *10 Hz* estimator bandwidth,
- a first order integrator,
- a *100 Hz* sample frequency.

5.2.5. Coulomb Friction Compensation

A simple compensation scheme for the nonlinear Coulomb friction is used:

$$T_{t1} = T_1 + 0.9\, F_{m1}\, sgn(\dot{\Phi}_1) \quad \text{(motor 1)} \tag{38}$$

$$T_{t2} = T_2 + 0.9\, F_{m2}\, sgn(\dot{\theta}_2) \quad \text{(motor 2)} \tag{39}$$

where T_{t1} is the total control torque 1
T_{t2} is the total control torque 2
and T_1 and T_2 are defined in (36)

Care must be taken in order to avoid overcompensation since this may cause stability problems [47]. Therefore only 90 % of the estimated Coulomb friction is compensated. The main effect of this compensation observed in the experiments is that the steady state error of the tip position is reduced by a factor 8.

5.3. Experimental Results

The experimental results of the controller presented above, are compared to the performance of a controller based on nonlinear decoupling for a rigid robot. The desired motion is a 0.858 m straight line in the *xy* plane starting from a point where the manipulator is almost completely extended. The motion has to be executed in 1.0 seconds.

5.3.1. Rigid Control

The nonlinear control law is obtained using the dynamic model for a robot without joint flexibility:

$$M_{11}(\theta_2)\begin{bmatrix}\ddot{\theta}_1\\ \ddot{\theta}_2\end{bmatrix} + \begin{bmatrix}F_{\theta_1}(\theta_1,\theta_2,\dot{\theta}_1,\dot{\theta}_2)\\ F_{\theta_2}(\theta_1,\theta_2,\dot{\theta}_1,\dot{\theta}_2)\end{bmatrix} = \begin{bmatrix}T_1\\ T_2\end{bmatrix} \tag{40}$$

with

$$F_{\theta_1} = C_{m1}\dot{\theta}_1 - 2\,M_c\,\dot{\theta}_1\dot{\theta}_2 sin(\theta_2) - M_c\,\dot{\theta}_2^2 sin(\theta_2)$$
$$F_{\theta_2} = C_{m2}\dot{\theta}_2 + M_c\,\dot{\theta}_1^2 sin(\theta_2)$$

The model is obtained from (35) by considering only the rigid part of the model, and by setting the motor position Φ_1 equal to the beam position θ_1. The control torque reduces to:

$$\begin{bmatrix}T_1\\ T_2\end{bmatrix} = \begin{bmatrix}F_{\theta_1}(\theta_{1d},\theta_{2d},\dot{\theta}_{1d},\dot{\theta}_{2d})\\ F_{\theta_2}(\theta_{1d},\theta_{2d},\dot{\theta}_{1d},\dot{\theta}_{2d})\end{bmatrix} + M_{11}(\theta_{2d})\begin{bmatrix}T_{fb1}\\ T_{fb2}\end{bmatrix}$$

$$T_{fb1} = \ddot{\theta}_{1d} - K_{\theta_1}(\theta_1 - \theta_{1d}) - K_{\dot{\theta}_1}(\dot{\theta}_1 - \dot{\theta}_{1d})$$

$$T_{fb2} = \ddot{\theta}_{2d} - K_{\theta_2}(\theta_2 - \theta_{2d}) - K_{\dot{\theta}_2}(\dot{\theta}_2 - \dot{\theta}_{2d})$$

The best results are obtained using a *1.5 Hz* controller bandwith, while the lowest resonance frequency is *4.6 Hz*. Figures 9a and 9b show the results.

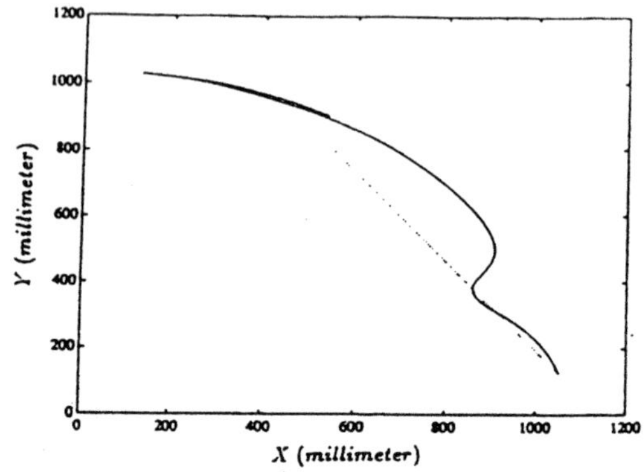

Figure 9a: Rigid control
The tip position trajectory (solid line)
compared with the desired trajectory (dashed line)

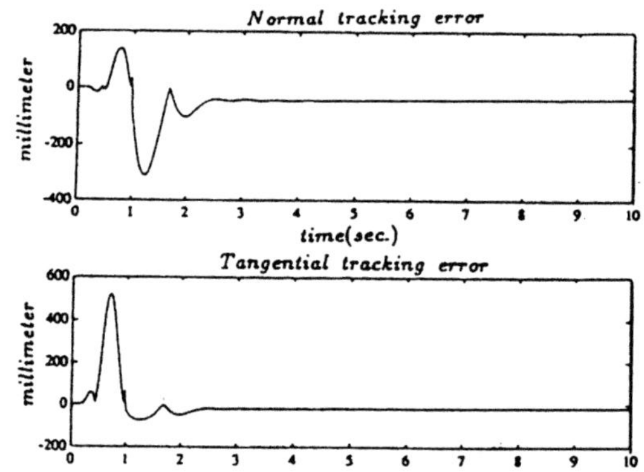

Figure 9b: Rigid control
Time history of the tracking error normal (top) and tangential (bottom) to the desired trajectory

The tip position trajectory in the *x-y* plane is compared with the desired trajectory. In addition, a time history of the tip position error normal and tangential to the desired trajectory is plotted. The maximum tracking error normal to the desired trajectory is *300 mm* while the error tangential to the desired trajectory is more than *500 mm*. The reason of these large tracking errors is that the large deflection of the springs is not taken into account. The steady state error normal to the desired trajectory is *35 mm* and the steady state error tangential to the desired trajectory is *18 mm*. The settling time is *3 seconds*.

5.3.2. GENERALIZED NONLINEAR DECOUPLING

The feedback gains are computed with respect to the final desired position. The best results are obtained with the poles placed at

$$\begin{bmatrix} -1.91 \\ -2.33 \\ -2.75 \pm 2.75\, i \\ -3.18 \pm 3.18\, i \end{bmatrix} \; Hz$$

Figures 10a and 10b show the results.

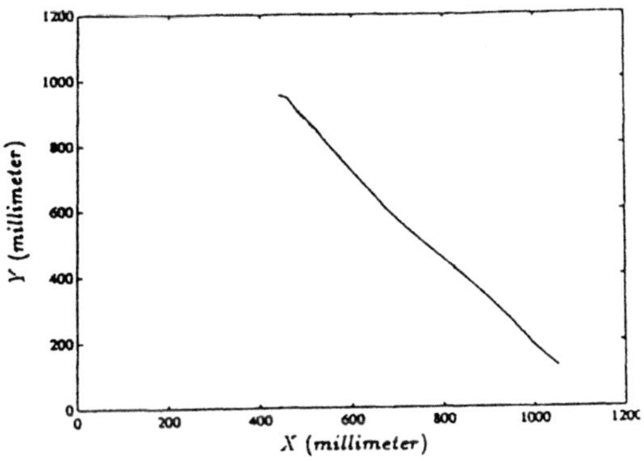

Figure 10a: Generalized nonlinear decoupling
The tip position trajectory (solid line)
compared with the desired trajectory (dashed line)

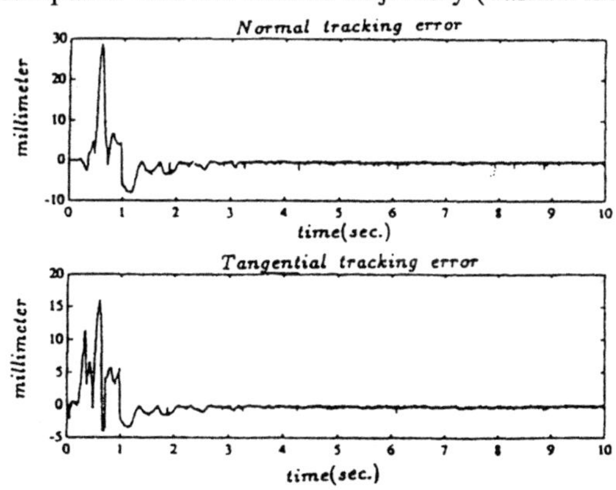

Figure 10b: Generalized nonlinear decoupling
Time history of the tracking error normal (top) and tangential (bottom) to the desired trajectory

After *1.3 seconds* the error is less than *2 mm*. Comparison with figures 9a and 9b reveals the superior performance of the method.

5.4. Conclusion

A method consisting of nonlinear feedforward to decouple and linearize the rigid body dynamics followed by linear feedback of the full system state is experimentally verified on a two link robot with flexible first joint.

Good control performance is obtained after:

- an accurate identification of the nonlinear model,
- inclusion of the nonlinear Coulomb friction,
- trajectory generation which takes into account the flexibility,
- use of a nonlinear discrete state estimator.

Comparison with a controller designed for rigid robots shows the superior behaviour of the controller.

6. Conclusion

This chapter has discussed the development and the experimental validation of continuous path tracking controllers for one-link and two-link flexible robots.

The design of the presented controllers involves experimental identification, feedforward and state feedback gains calculation.

The models, on which the control design is based, take into account part of the robot flexibility. The feedforward takes care of ensuring good trajectory tracking properties. Its calculation is based on the inverse robot model. State feedback eliminates small tracking errors due to environmental disturbances and modelling errors. The signals of extra noncollocated sensors, which measure the deformation of the flexibility, are in addition to the normal collocated sensor signals used for the state feedback.

The tests show that these controllers are superior to conventional PID-based robot controllers in terms of overshoot, settling time and tracking accuracy. Their industrial application is expected to reduce the production cycle time and to improve the production accuracy significantly. This should then result in lower production costs and in an improvement of the end product quality.

Bibliography

[1] Kuntze, H.B., Jacubasch, A.H.: Control algorithms for stiffening an elastic industrial robot. IEEE journal of Robotics and Automation, Vol. RA-1, N°2, pp 71-78, June 1985.

[2] Gerung, M.: Accuracy measurement on a KUKA IR 161/60 with an RCM3. Technical Report, Esprit project 1561: Sacody, 1990.

[3] Swevers, J., Torfs, D., Adams, M., De Schutter, J., Van Brussel, H.: Comparison of control algorithms for flexible robots implemented on a KUKA IR 161/60 industrial robot. Proceedings of the Fifth International Conference on Advanced Robotics, Italy, 19-22 June 1991.

[4] Cannon, R.H., Schmitz, E.: Initial experiments on the end-point control of a flexible one-link robot. International Journal of Robotic Research, Vol. 3, N° 3, pp 62-77, Fall 1984.

[5] Rovner, D.M., Cannon, R.H.: Experiments towards on-line identification and control of a very flexible one-link manipulator. International Journal of Robotic Research, Vol. 6, N° 4, pp 3-19, Winter 1987.

[6] Uhlik,C.R.: Experiments in high performance nonlinear and adaptive control of a two-link flexible-drive-train manipulator. Ph.D. Thesis, Stanford University, Department of Aeronautics and Astronautics, Stanford, California, May 1990.

[7] Swevers, J., Torfs, D., De Schutter, J., Van Brussel, H.: Accurate tracking control of a flexible one-link robot. Proceedings of the 15th International Seminar on Modal Analysis, Leuven, Belgium, 19-21 September 1990, pp 353-366.

[8] Pfeiffer, F., Gebler, B.: A multistage approach to the dynamics and control of elastic robots. Proceedings of the IEEE conference on Robotics and Automation, Philadelphia, 1988, pp 2-8.

[9] Adams, M., De Schutter, J.: Experiments on nonlinear control of flexible robots. Proceedings of the 15th International Seminar on Modal Analysis, Leuven, Belgium, 19-21 September 1990, pp 635-648.

[10] Spong, M.W.: Modeling and control of elstic joint robots. Transactions of the ASME, Vol. 109, December 1987, pp 310-319.

[11] Demeester, F., Van Brussel, H.: Real time optical measurements of robot structural deflections. Mechatronics, Vol.1, N°1, 1991, pp 73-86.

[12] Swevers, J., Torfs, D., Demeester, F., Van Brussel, H.: Fast and accurate tracking control of a flexible one-link robot based on real-time link deflection measurements. Accepted for publication in Mechatronics

[13] Van Der Auweraer, H.: Off-line identification of a KUKA IR 161/60 industrial robot. Technical report, task 4851, Esprit project 1561: Sacody, 1990.

[14] Widmann, G.R., Ahmad, S.: Control of industrial robots with flexible joints. Proc. IEEE Conf. on Robotics and Automation, pp 1561-1566, 1987.

[15] Rivin, E.I.: Effective rigidity of robot structures: analysis and enhancement. Proc. of the 1985 American Control Conf., Boston pp 381-382, June 1985.

[16] Futami, S., Kyura, N., Hara, S.: Vibration absorption control of industrial robots by acceleration feedback. IEEE Transactions on Industrial Electronics, Vol. IE-30, No.3, pp 299-305, August 1983.

[17] Hollars, W.G.: Experiments in end-point control of manipulators with elastic drives. Ph.D. Thesis, Stanford University, Stanford, California, Department of Aeronautics and Astronautics, May 1988.

[18] Adams, M., Swevers, J., Torfs, D., De Schutter, J., Van Brussel, H.: Implementation of control algorithms for flexible joint robots on a KUKA IR 161/60 industrial robot: experimental results. Proc. of the Second International Symposium on Experimental Robotics, Toulouse, France, 25-27 June 1991.

[19] Book, W.J., Majette, M.: Controller design for flexible, distributed parameter mechanical arms via combined state space and frequency domain techniques.J. of Dynamic Systems, Measurements, and Control, Vol. 105, pp 245-254, December 1983.

[20] Hastings, G.G., Book, W.J.: A linear dynamic model for flexible robotic manipulators, IEEE Control Systems Magazine, pp 61-64, February 1987.

[21] Gebler, B.: Feed-forward control strategy for an industrial robot with elastic links and joints. Proc. IEEE Conf. on Robotics and Automation, pp 923-928, 1987.

[22] Shung, I.Y., Vidyasagar, M.: Control of a flexible robot arm with bounded input optimum step responses. Proc. IEEE Conf. on Robotics and Automation, pp 916-923, 1987.

[23] Wang, D., Vidyasagar, M.: Control of a flexible beam for optimum step response. Proc. IEEE Conf. on Robotics and Automation, pp 1567-1572, 1987.

[24] Korolov, V.V., Chen, Y.H.: Controller design robust to frequency variation in a one-link flexible robot arm. J. of Dynamics Systems, Measurements, and Control, Vol. 11, pp 9-14, March 1989.

[25] Wells, R.L., Schueller, J.K., Tlusty, J.: Feedforward and feedback control of a flexible robotic arm. IEEE Control Systems Magazine, pp 9-15, January 1990.

[26] Devolder, P., Devolder, P.: "Online" identificatie en adaptieve controle van een flexibele robot met twee vrijheidsgraden. Ingenieursthesis, Faculteit Toeg. Wetenschappen, K.U. Leuven, 1990.

[27] De Schutter, J., Van Brussel, H., Adams, M., Froment, A., Faillot, J.L.: Control of flexible robot using generalized nonlinear decoupling. Proc. Symposium on Robot control, Karlsruhe pp 98.1-98.6, 1988.

[28] De Jaegere, P., Van Brussel, H.: The extension of a load compensation method to control of flexible joint robots submitted to Int. J of Robotics Research.

[29] Tomizuka, M.: Zero phase error tracking algorithm for digital control. J. of Dynamic Systems, Measurement and Control, Vol. 109, pp 65-68, March 1987.

[30] Torfs, D., Swevers,J., De Schutter, J.: Extended bandwidth zero phase error tracking control of non-minimal phase systems submitted to J. of Dynamic Systems, Measurements and Control.

[31] Swevers, J., Torfs, D., De Schutter, J., Van Brussel H.: Experiments on fast and accurate tracking control of a flexible one-link robot. Presented at the 1991 IEEE Conf. on Robotics and Automation, and internal report 91P21, K.U. Leuven, Department of Mechanical Engineering, 1991.

[32] Swevers, J., Adams, M., De Schutter, J., Van Brussel, H., Thielemans, H.: Limitations of linear identification and control techniques for flexible robots with nonlinear joint friction. Experimental Robotics 1, Lecture Notes in Control and Information Science, Springer-Verlag, 1990.

[33] Asada, H., Kanade, T., Takeyama, I.: Control of a direct drive arm. J. of Dynamic Systems, Measurement and Control, vol. 105, pp 136-142, September, 1983.

[34] Torfs,D., Swevers, J., De Schutter J.: Quasi-perfect tracking control of nonminimal phase systems presented at the 1991 IEEE Conf. on Robotics and Automation, and internal report 91P22, K.U.Leuven, Department of Mechanical Engineering, 1991.

[35] Timoshenko, S., Young, D., Weaver, W.: Vibration problems in engineering. John Wiley & Sons, New York, 1974.

[36] Swevers,J., De Moor, B., Van Brussel, H.: Stepped sine system identification, errors-in-variables, and the quotient singular value decomposition. Proc. of the 15th International Seminar on Modal Analysis, Leuven , Belgium, pp 59-76, 19-21 September 1990, and Mechanical Systems and Signal Processing Vol. 5, No. 6, November 1991.

[37] Swevers, J.: An introduction to system identification and its application in practice. Computer Controlled Motion and Robotics Course, K.U. Leuven, Belgium, 1991.

[38] Faillot, J.L.: Examine state of the art of control of flexible structures. Technical report, task 3121, Esprit Project 1561: Sacody, 1987.

[39] Duque, M., Samaan, M.: Partial state LQ and GPC adaptive control: an experimental evaluation. Proc. of the Eighth International Conference on Analysis and Optimization of Systems, Springer-Verlag, Antibes, France, pp 830-840, June 1988.

[40] Freund. E.: Fast nonlinear control with arbitrary pole-placement for industrial robots and manipulators. Int. J. of Robotics Research, Vol. 1, pp 65-78, 1982.

[41] Bortoff, S.A., Spong. M.W.: Feedback linearization of flexible joint robot manipulators. Proc. 26th Conf. on Decision and Control, Los Angeles, pp 1357-1362, 1987.

[42] Forrest-Barlach, M.G. Babcock, S.C.:Inverse dynamic position control of a compliant manipulator. IEEE J. of Robotics and Automation, Vol. RA-3, No 1, pp 75-83, 1987.

[43] De Luca, A., Isidori, A., Nicolo, F.: An application of nonlinear model matching to the dynamic control of robot arm with elastic joints via nonlinear dynamic feedback. Proc. IEEE 24th Conf. on Decision and Control, Ft. Lauderdale, FL, pp 1671-1679, 1985.

[44] Singh, S.N., Schy, A.A.: Robust torque control of an elastic robotic arm based on invertibility and feedback stabilization. Proc. IEEE 24 th Conf. on Decision and Control, Ft Lauderdale, FL, 1985.

[45] Hollerbach, J.M.: Dynamic scaling of manipulator trajectories. ASME Journal of Dynamic Systems, Measurement and Control, Vol. 106, pp 102-106, 1984.

[46] Van Den Bossche, J.: RODYM : A new approach to robot metrology. Internal report, K.U. Leuven, Department of Mechanical Engineering, Leuven, Belgium, 1990.

[47] Kubo, T., Anwar, G., Tomizuka, M.: Application of nonlinear friction compensation to robot arm control. IEEE Proc. Conf on Robotics and Automation, pp 722-726, 1986.

PART II

A Toolbox for Lightweight Robot Design and Verification

Chapter 3

On the Identification of the Dynamic Characteristics of Flexible Robots

H. Van der Auweraer
LMS International

1. Objectives

The objectives of the identification related tasks performed in the SACODY project are the study and implementation of methods for off-line measurement, identification and interpretation of the dynamic robot characteristics.

This detailed knowledge of the robot dynamic properties is required for the optimization of the robot design as well as for the design of an optimal controller which must take these dynamic characteristics into account. In addition to providing the necessary evaluation information, the experimental dynamic analysis results are also used to optimize and update theoretical models.

Such models are essential to simulate the dynamic and kinematic behaviour of the robot system, including the controller and actuators. By numerical simulation, the robot performance can be evaluated from in the design phase, while the effect of the controller system can be predicted.

2. Overview

To realise these objectives, a critical review of the state of the art of identification methods was made. Based on the conclusions of this review, two particularly interesting identification techniques were further developed, one using a direct time domain identification approach, and one using a frequency domain direct parameter identification approach. In both cases, a state space model formulation is used.

The frequency domain technique was integrated in a commercial system for Computer Aided Dynamic Analysis while the time domain technique was made available as a prototype software.

The developed techniques have been applied to the identification of the studied breadboard systems as well as to the dynamic analysis and modelling of the two investigated robots (KUKA IR-360 and KUKA IR-160).

Since, in general, system identification techniques use linear system models, the existence and influence of system nonlinearities has also been investigated. Specific tools for detecting and describing system nonlinearities have been developed and applied to the study of robot and breadboard systems.

The experimental identification data can be used to update the numerical model of the system under study. This updated model can then be used to describe the flexible components of the multibody dynamical system. In order to use the identification results for simulation purposes, an interface was developed to calculate the necessary complementary modal moment information and to export the dynamic model data to the simulation software package.

3. Review of Identification Methods [1]

Most currently available identification techniques for structural analysis purposes identify a modal system model in terms of the structure modal parameters: natural frequencies, damping ratios and mode shapes. These parameters are in general estimated indirectly, from the measured impulse or frequency response functions or from the structure free decay response.

Recent research efforts in this area have concentrated on improving the accuracy and consistency of the obtained models for systems with a large number of degrees of freedom. The most advanced of these techniques estimate models which describe the measurements as well as possible in a global least squares sense: measurements for all responses and inputs are analysed simultaneously in order to yield a unique, global, model for them.

The most popular of these indirect modal techniques are:

- Least Squares Complex Exponential or Polyreference method, estimating the parameters of damped complex exponentials to describe the impulse responses. Via the system's covariance matrix, the coefficients of a companion matrix are estimated. The complex system poles are then obtained from its eigenvalue decomposition. A linear least squares fit is then used to estimate the mode shapes.
- Ibrahim Time Domain Method (ITD), using the free decay responses and their time-shifted versions to estimate a matrix whose eigenvalue and eigenvector decomposition yields the modal parameters.
- Rational Polynomial methods, estimating the matrix coefficients of a frequency domain equivalent of an auto-regressive moving average system model, which takes the form of a ratio of matrix polynomials. Via an eigenproblem analysis, the modal parameters can then be obtained.

Other techniques allow the direct usage of measured input-output data. In the Direct Parameter Time Domain technique, the matrix coeficients of the auto-regressive moving average finite difference equation are estimated directly from the measured time data. Again, an eigenvalue decomposition yields the required modal parameters.

In a totally different approach, a state-space model is estimated for the structure, based on measured input-output signals. An example of this class of methods is the Eigensystem Realisation algorithm (ERA). This method estimates a minimum order realisation for a block Hankel matrix of the measured signals. From the obtained state-space model, the state transition matrix is solved for eigenvalues and -vectors.

Since all of these techniques analyse all available measurement data simultaneously, substantial computer calculation speed, accuracy and memory capacity are required. This is especially true for the ERA-algorithm based on a singular value decomposition of the complex data matrix.

As far as these aspects are concerned, the Polyreference method is by far preferable, since the dimension of the matrices is related to the (limited) number of inputs, rather than to the number of outputs as for all other techniques. Moreover, the concept of the covariance matrix also limits the memory load, and the canonical form of the used model drastically reduces the number of unknowns to be solved for.

Unfortunately, this technique does not allow to derive the system's state space model directly. Another important disadvantage, which holds for all time-domain techniques, is the lack of capabilities to correct for out-of-band effects when only a small number of modes has to be identified. This is typically the case for the control of flexible robot structures where only the first few modes of vibration significantly influence its positioning accuracy. In this identification, the contribution of relatively highly damped modes outside the narrow frequency band of interest might be important and can only be modelled correctly by frequency domain methods.

Another important element in this discussion concerns the nature of the data to start the identification from. In the indirect approaches, the input-output time data are first reduced to a non-parametrical system description such as impulse responses or FRFs. By applying specific testing techniques such as stepped sine testing and by using data averaging procedures, the quality of the functional model can be substantially improved without increasing the total amount of data. This is contrary to the direct approaches, where the raw time data are directly processed. Reducing the uncertainty on the results can only be obtained through the processing of extended data sequences, which leads to a highly increased estimation duration.

Based on these conclusions it was decided to further investigate two approaches in the context of the identification of state-space models for flexible structures.

The first approach consists of a frequency domain technique, capable of deriving a state-space model in terms of generalised states. Since the method can (although not necessarily has to) follow the indirect approach, it is well adapted to the analysis of complex (possibly nonlinear) systems using high-quality FRF estimation techniques such as stepped sine excitation. This is especially important when the excitation is applied through the actuator system.

Specific condensation techniques have been developed to reduce the equation dimensions to the effective rank of the dataset. Residual effects of modes outside the frequency range of analysis can explicitly be taken into account, not only for the mode shape estimation, but also in the pole calculation phase.

This approach is specifically suited to perform a full-scale dynamic identification of the complete robot structure.

The second approach consists of the further development of a direct time domain identification method which is capable to generate a state-space system model, directly using the input-output data. Special attention was paid to overcome the traditional drawbacks of direct time domain identification methods. The method is specifically suited to identify, in a very direct way, reduced models of the breadboard systems and of the first, critical, modes of the robot structure.

4. Development and Implementation of Parameter Identification Methods

As discussed above, two identification approaches were further developed. The frequency domain method was primarily developed by LMS and integrated in the LMS software for Computer Aided Dynamic Analysis. The method was applied to the identification and evaluation of the basic structural and servo/structural behaviour of the robots, and its results were applied to develop valid simulation models.

The time domain approach was primarily studied at the K.U.Leuven and was implemented as a prototype software used in the control of the breadboard and robot systems.

4.1. Frequency Domain Identification [2,3,4]

The developed frequency domain parameter identification technique allows to identify the parameters of both a state space and a modal system model from measured frequency response function (FRF) data as well as from measured input-output spectra. A proper selection of the state parameters enables to identify the model matrices using the constitutive system equations. The actual implementation of the technique is made very effective, using a condensation of the multi-point FRF data into a set reflecting the effective number of degrees of freedom.

4.1.1. State Space Formulation

The basis for the frequency domain direct parameter identification procedure is the state-space representation of the input-output relations for a mechanical system:

$$\dot{x}(t) = Ax(t) + Bu(t)$$
$$y(t) = Cx(t)$$

with:

- $x(t)$: state variable ($2N$ elements)
- $y(t)$: output signal (N_o elements)
- $u(t)$: input signal (N_i elements)
- A : state transition matrix ($2N$, $2N$)
- B : input matrix ($2N$, N_i)
- C : output matrix (N_o, $2N$)

Similarly, a Laplace domain equivalent of this formulation can be derived (assuming the system starts from rest, $\dot{x}(0) = 0$):

$$sX(s) = AX(s) + BU(s)$$

$$Y(s) = CX(s)$$

By solving for the state variable $X(s)$, the following input-output relation is obtained,

$$X(s) = [sI-A]^{-1} BU(s)$$

corresponding to a Transfer Function formulation,

$$H(s) = C[sI-A]^{-1}B$$

For this system the corresponding impulse responses $h(t)$ are thus defined by:

$$h(t) = Ce^{At}B$$

The state variable $x(t)$ in this formulation is general. However, a proper choice of the system state variables allows to relate above state space formulation with the system's constitutive equations.

4.1.2. Relation with Constitutive Equations

For a linear mechanical system with N degrees of freedom and N_i inputs, the relationship between the displacements $r(t)$ resulting from a force input $f(t)$ (both defined as a vector for the N degrees of freedom), is given by a set of N second order differential equations:

$$Ma(t) + Di(t) + Kr(t) = f(t)$$

with

- $a(t)$: the acceleration response vector
- $i(t)$: the velocity response vector
- M : the mass matrix
- D : the damping matrix
- K : the stiffness matrix

Assuming a non-singular mass matrix M, above equation can be reformulated as:

$$a(t) = -A_1 i(t) - A_0 r(t) + M^{-1}f(t)$$
$$i(t) = I\, i(t)$$

with

$$A_1 = M^{-1}D$$
$$A_0 = M^{-1}K$$

When now selecting the state variable $x(t)$ in the state space equation such that it is composed of a displacement $r(t)$ and a velocity $i(t)$ fraction, each of dimension N, it is readily seen that the system matrices A, B and C take the following form:

$$A = \begin{bmatrix} -A_1 & -A_0 \\ I & 0 \end{bmatrix}$$

$$B = \begin{bmatrix} M^{-1} \\ 0 \end{bmatrix}$$

$$C = \begin{bmatrix} 0 & I \end{bmatrix}$$

This implies that, when the matrices A_1, A_0 and M can be identified, at the same time a state space system model will be available.

4.1.3. Matrix Identification

The actual identification of the different matrices is done in the Laplace domain, using measured frequency response functions. To this purpose, the constitutive equations in the time domain are first rewritten considering the response of the system to Dirac impulses. Considering an N by Ni force matrix $f_d(t)$ which is composed in such a way that all columns contain zeros except at one location where a Dirac impulse represents the force, the constitutive equation set takes the following form:

$$\ddot{h}(t) + A_1 \dot{h}(t) + A_0 h(t) = M^{-1} f_d(t)$$

with

$h(t)$: the impulse response function matrix
$\dot{h}(t)$: the first derivative of the impulse response matrix
$\ddot{h}(t)$: the second derivative of the impulse response matrix
In the Laplace domain, the equivalent of these derivatives is:

$$\begin{aligned} \mathcal{L}[h(t)] &= H(s) \\ \mathcal{L}[\dot{h}(t)] &= s\,H(s) - h(0) \\ \mathcal{L}[\ddot{h}(t)] &= s^2\,H(s) - s\,h(0) - \dot{h}(0) \end{aligned}$$

or, by using the state space expression for the impulse response function, and taking into account that $C.B = 0$ because of the state variable definition,

$$h(t) = C\,e^{At}\,B$$

$$\begin{aligned} \mathcal{L}[h(t)] &= H(s) \\ \mathcal{L}[\dot{h}(t)] &= s\,H(s) - CB = s\,H(s) \\ \mathcal{L}[\ddot{h}(t)] &= s^2\,H(s) - s\,CB - CAB = s^2\,H(s) - CAB = s^2\,H(s) - M^{-1} \end{aligned}$$

The constitutive equation set now reduces to:

$$[s^2 + s\,A_1 + A_0]\;H(s) = 2\,M^{-1} = B_0$$

This equation means that the measurement data set can be described by a second order linear model with real constant matrices. Since this equation is valid for each value of the Laplace variable, and in particular along the frequency axis $s = jw$, all these equations can be gathered in an overdetermined set which can be solved in a global least squares sense for the unknown system matrices. From these matrices, the state space system matrices are directly derived.

From the system matrix A, an eigenvalue decomposition yields the system poles and the modal matrix of mode shapes [2, 3].

4.1.4. Data Condensation

Since, in general, only a limited number of system modes have to be identified using data from a large number of response stations, the coefficient matrices in the system of equations become rank deficient. On the other hand, the system matrices are of large dimensions and the equation size is large, with all its implications on computer calculation speed and memory requirements.

Therefore, a data condensation procedure was implemented that reduces the measured data using a proper basis for the space that is spanned by these data. The actual dimension of this space is an important parameter in order not to loose any information. This value, which represents the number of effective degrees of freedom, is determined from a singular value analysis of the measured data.

The transformation matrix T which describes the actual condensation by

$$H'(s) = TH(s)$$

is then constructed from the singular vectors corresponding to the selected number of largest singular values. As a result, a set of reduced FRFs and a representative state $X'(s)$ and the corresponding output matrix C' can be identified. The actual parameter identification, including the modal decompostion, then takes place using the reduced set of FRFs.

The poles and modal participation vectors are the same as for the uncondensed model. The modeshapes however are expressed in the reduced state basis. The original state vector $X(s)$ and the full mode shapes can now easily be obtained by applying the corresponding backtransformation given by T^T.

4.1.5. Characteristics

The described identification technique has the following important characteristics:

- The state transition matrix A and input and output matrices B and C are readily available.
- A full modal parameter model is derived by an eigenvalue analysis.

- A very important problem, namely the dimension of the state space model, is eliminated by the calculation of the reduced state via the "singular" value analysis of the data of the measurement matrix.
- The number of unknowns is limited, using only a second order linear model with matrix coefficients of a dimension N_e, the number of effective degrees of freedom.
- Numerically reliable tools are used: least squares solution and eigenvalue decomposition techniques.
- All measurement data can be analysed simultaneously, reducing the measurement noise, and yielding a consistent model for the system's response.
- Being a frequency domain method, residual effects of modes outside the frequency band of analysis, can explicitly be taken into account, not only for the modeshape estimation, but even for the system matrix and pole calculation.
- The method is compatible with high-quality testing techniques such as FRF based stepped sine testing.

4.2 - Direct Time Domain Identification [4,5,6]

Direct time domain identification methods derive a state space model for multivariate linear time invariant systems directly from input-output measurements. A priori information about the system is very restricted. The starting point of these methods is a structured input-output relationship, derived from the state space model.

Most of the available direct identification methods select linear independent rows of the structured matrices of the input-output relations. This has two important drawbacks. First, this takes many rank tests of large matrices. Secondly, noise perturbated data always lead to a situation where all rows are linearly independent, although this does not reflect the theoretical system case, formulated for noise-free data.

Besides, these direct identification methods very quickly degenerate in purely algebraic manupulations of formulas, without getting insight into the problem of the identification.

Therefore, a new approach was pursued, also starting from a structured input-output relationship, but avoiding the difficulties with the selection of linearly independent rows. Instead the properties of block Hankel matrices and the numerically very attractive technique of singular value analysis are used. The investigated methods turn out to be simple and geometrically interpretable.

4.2.1. Hankel Matrix Method

The discrete time state space model of an N-dimensional system is given by:

$$x_{k+1} = Ax_k + Bu_k$$

$$y_k = Cx_k + Du_k$$

with A, B, C, D real matrices, and the vectors u_k, y_k and x_k respectively the input-, output- and state vectors.

The basis for the method is a structured input-output description of the following form:

$$Y_h(k,i,j) = G(i)X(k,j) + H_t(i)U_h(k,i,j)$$

with,

$Y_h(k,i,j)$ a block Hankel matrix constructed by the $i+j-1$ consecutive output vectors, starting with y_k:

$$\begin{bmatrix} y_k & y_{k+1} & y_{k+2} \cdots\cdots y_{k+j-1} \\ y_{k+1} & y_{k+2} & y_{k+3} \cdots\cdots y_{k+j} \\ & & \\ y_{k+i-1} & y_{k+i} & y_{k+i+1} \cdots\cdots y_{k+i+j-2} \end{bmatrix}$$

$U_h(k,i,j)$ has the same structure as $Y_h(k,i,j)$, but is constructed using the input vectors. $G(i)$ is an extended observability matrix:

$$\begin{bmatrix} C \\ CA \\ CA^2 \\ \cdots \\ \cdots \\ CA^{i-1} \end{bmatrix}$$

and $X(k,j)$ a concatenated state matrix: $[x_k\, x_{k+1}\, x_{k+2} \dots x_{k+j-1}]$

and $H_t(i)$ a block Toeplitz matrix with the first i Markov parameters, given by

$$H_i = CA^{i-1}B$$

$$\begin{bmatrix} H_0 & 0 & 0 \cdots\cdots 0 \\ H_1 & H_0 & 0 \cdots\cdots 0 \\ H_2 & H_1 & H_0 \cdots\cdots 0 \\ H_{i-1} & H_{i-2} & H_{i-3} \cdots\cdots H_0 \end{bmatrix}$$

The actual identification procedure then uses the concatenated matrix

$$\begin{bmatrix} Y_n(k,i,j) \\ U_n(k,i,j) \end{bmatrix}$$

For the estimation of the actual system order and the computation of the matrices A, B, C, D, the left singular vectors from the concatenated matrix need to be calculated [4].

The concept of singular value analysis allows to deal appropriately with the case where noise is present on the input-output data. In practice, limits have to be specified for the lowest acceptable singular values, which are attributed to the system behaviour. It can be proven that, with increasing overdetermination of the equation set, the singular values that are used to determine the actual order of the observable part of the system, grow towards certain asymptotic levels, allowing to distinguish clearly the system singular values from those caused by the noise.

In order to calculate the singular value decomposition (S.V.D) of a largely overdetermined concatenated matrix, a two-step procedure is used. First a RQ factorisation is computed, followed by the S.V.D of the R-part.

4.2.2. Canonical Correlation Method

The Hankel matrix method was further improved to make it more straightforward and less computationally involved. The method basically consists of two steps.

First, a state vector sequence is realised as the intersection of two row spaces of two block Hankel matrices, constructed with input-output data.

Then the system matrices are obtained at once from the least squares solution of a set of linear equations. A computationally efficient way of computing the system matrices was implemented, making use of available singular value decompositions realised in the previous step.

The time domain method was implemented on the prototype identification system developed at the K.U.Leuven. It was applied to the identification and control of the breadboard systems, as discussed in Section 2 on the toolbox justification.

4.3. Integration with Simulation [7,8,9]

One of the main objectives of the off-line identification of a full dynamic model of the robot structures is to enable the optimisation of numerical models which can be used for simulating the robot behaviour for various loading and controller configurations. Such simulation software is provided by the ADAMEUS software, described elsewhere in the report.

The simulation model used by ADAMEUS to describe the dynamics of a flexible system is based on the modal characteristics of the individual deformable bodies. These flexible properties were up to now only available from a modal model obtained in a theoretical way using the finite element method. One of the objectives of the SACODY project is to introduce in ADAMEUS flexible properties resulting from both experimental and analytical modal analysis data.

The modal synthesis method of the simulation package ADAMEUS defines the flexibility of a deformable body by a number of modes. The data associated with each mode and necessary to characterise a deformable body are:

- the generalised mass,
- the generalised stiffness or the natural frequency,
- the mode shape coefficients in a number of nodes: the components interconnection nodes and the grids for actuators etc.,
- the linear momentum of the modal motion at the center of mass,
- the kinetic momentum of the relative rotational motion of the modes.

The former three data types are generally available from any modal identification procedure. The latter two however require the knowledge of the mass distribution. This information is only available from finite element calculation. Starting from the element mass matrices from the FEM analysis, a lumping scheme must be executed to distribute concentrated masses on the finite element nodes.

Several mass lumping techniques were evaluated and tested [8]. The selected method, proposed by Surana, uses a lumping scheme which takes the consistent mass and scales diagonal terms off, thus preserving the total mass.

4.4. Integration in CAE Software

The developed frequency domain direct parameter identification technique has been integrated into the LMS CADA-X software for Computer Aided Dynamic Analysis.

The identification technique was integrated in the CADA-MODAL product for modal analysis. It can be selected as one of the available methods for single reference as well as for multiple reference parameter identification. In both cases, the method acts as a global estimator, processing all responses simultaneously.

For the definition of the system geometry, the visualisation of animated mode shapes, the comparison of modelled with measured data and various other tools for optimal order selection and result validation, the standard available CADA-MODAL functionalities are used. All user interface and graphical formats, data representation and data storage are fully compatible with the rest of the CADA-X software.

The procedure for calculating the kinetic moments, required for the simulation of the dynamics of multibody flexible systems, was integrated as a data preprocessor in the CADA-LINK product.

This product was conceived to integrate and manipulate data from Test and from FEM for dynamic analysis and it features various functionalities for correlation and updating of numerical models using experimental data. The CADA-LINK database contains all necessary data to evaluate the flexible characteristics of a deformable body.

CADA-LINK also allows to exchange model data with external sources by means of a "neutral" interface, the format of which was defined in the Esprit Project 322, "CAD*I" [9]. An additional option in the writing of the "neutral file" was added, which enables the calculation and output of the necessary moment information in addition to the standard modal model data.

The calculated momentum information can be visualised by means of the standard graphical functionalities of the CADA-LINK product. The definition of the Neutral File layout was extended to include the momentum description [8].

5. Identification of Demonstration Robots

The developed frequency domain identification procedure was applied to the identification of two robot structures. In the beginning of the second project year, the identification of the KUKA IR-360 robot was performed as a preliminary evaluation of the dynamic behaviour of typical flexible structures. A first prototype version of the identification software was available by then and consequently applied.

At the end of the third project year, the actual identification of the selected demonstration robot, being a KUKA IR 161, was performed using the final integrated version of the identification method.

5.1. Identification of the KUKA IR-360 Robot

In order to evaluate the basic dynamic behaviour of the robot, as well as its dependency on the position of the various links, a modal parameter identification was performed in 5 different configurations [see Figure 1].

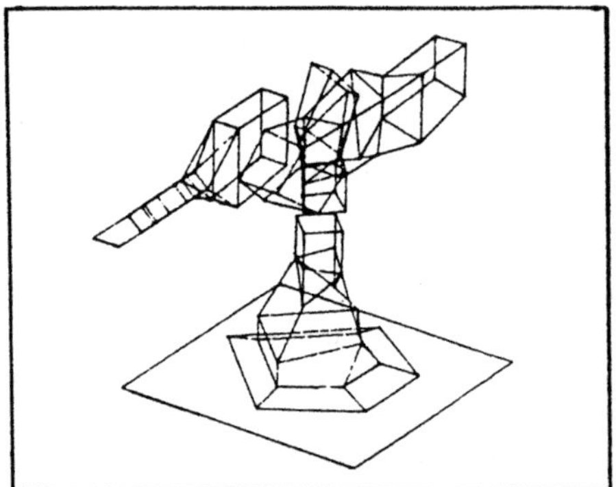

Figure 1a : IR 360, Configuration 1,2

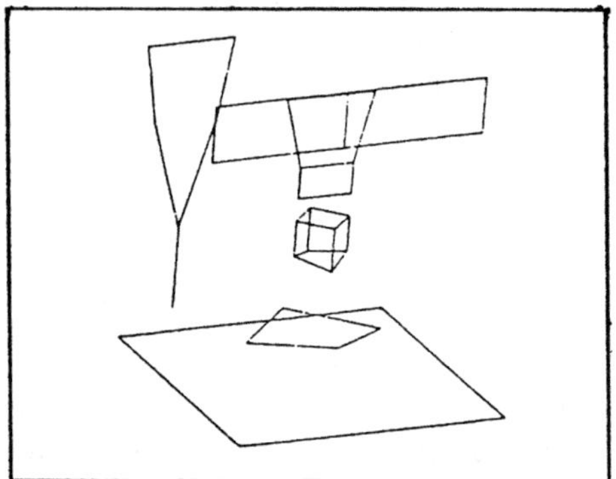

Figure 1b : IR360, Configuration 3

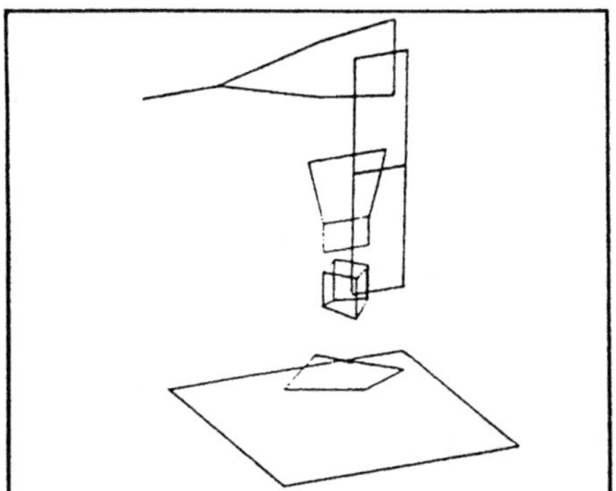

Figure 1c : IR360, Configuration 4

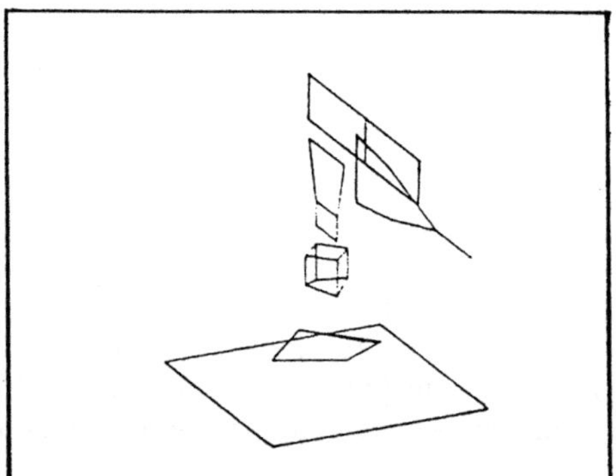

Figure 1d : IR360, Configuration 5

In the first configuration, an extended set of measurement locations was used. Since the fundamental modes of the robot appeared to be defined mainly by the local flexibility of the axis and drive system, and not by the link element flexibilities, the other configurations were measured in a reduced set of locations.

The second configuration is identical to the first one, the excitation level being three times higher.

All identifications were performed using single input excitation at an excentric location above the first axis-joint. Periodic random excitation was applied and the Hv FRF estimator was used. Special attention was paid to the overall test set-up, force spectrum realisation and data acquisition.

Below, the main results for the first configuration are presented. The complete discussion of all results is given in [10].

In the extended setup, 106 response locations were measured in three directions, resulting in a set of 318 FRFs. A condensed representation of this data set is given by the summed amplitude spectrum and the corresponding Mode Indicator function (lower and upper trace in Figure 2). A singular value analysis confirms that in the analysed frequency band, up to *30 Hz*, only 4 significant modes contribute to the dynamic behaviour.

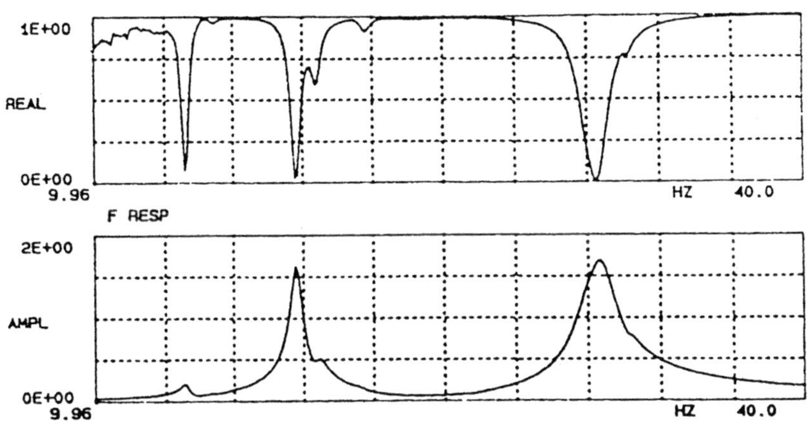

Figure 2 : Mode Indicator Function (upper), Summed FRF (lower)

All 318 FRFs were analysed simultaneously, yielding the following results:

Mode	Frequency	Description
1	*13.88 Hz*	Rotation of axis 1 around Z-axis
2	*18.68 Hz*	Tilting of the upper robot structure around the Y-axis
3	*19.62 Hz*	Tilting around the X-axis
4	*21.69 Hz*	Bending of the robot on its base, around the X-axis

Figure 3 shows the mode shapes of the first two modes on the 3D wire frame model.

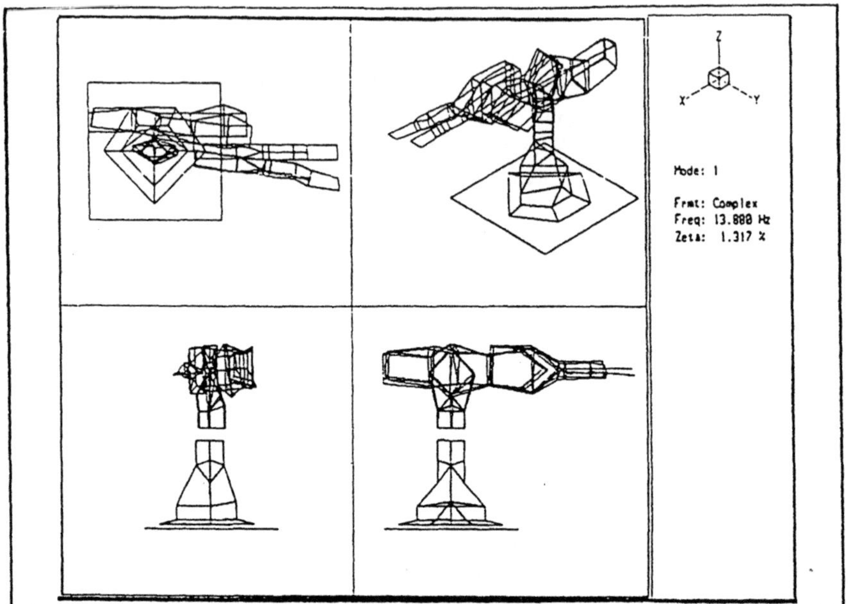

Figure 3a : IR360, Configuration 1 ; First Mode

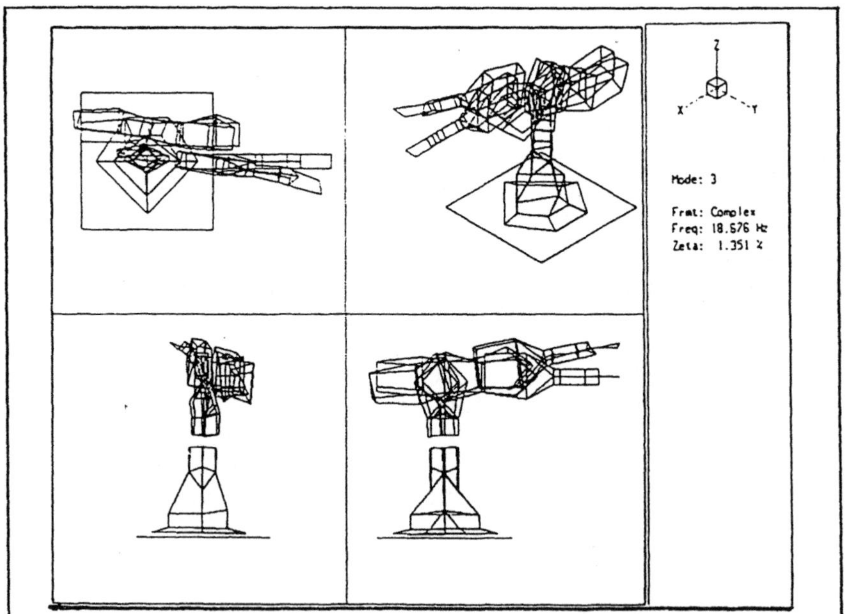

Figure 3b : IR360, Configuration 1 ; Second Mode

The same basic modes were encountered in all configurations, the frequency of the various modes is however dependent on the actual position of the links, even to an extent that the relative location of the resonances can vary with the configuration. For the resonance frequencies of the first three modes, which are articulated around the first joint, this gives the following summary *(in Hz)*:

Modeshape	Conf.1	Conf.2	Conf.3	Conf.4	Conf.5
Rotation around Z-axis	*13.88*	*13.96*	*14.98*	*23.46*	*14.25*
Tilting around Y-axis	*18.68*	*18.82*	*19.18*	*16.83*	*23.33*
Tilting around X-axis	*19.62*	*20.02*	*20.14*	*13.54*	*19.84*

5.2. Identification of the KUKA IR-160 Robot

The aim of this experimental identification was to determine the dynamic model for the KUKA IR 161 robot in the low frequency region.

The dynamic characterization is mainly achieved in two different ways. First, an experimental modal analysis was performed to describe the dynamic behaviour in terms of natural frequencies, damping values and modeshapes. These data are determined out of frequency response functions of acceleration over force. The force is externally generated with a vibration exciter.

Next, frequency response functions in terms of acceleration over motor current are measured. In this case the excitation is generated by each of the first three motors separately.

As the dynamic properties vary with the position of the robot, the analysis is carried out for different position configurations.

The modal analysis is carried out in three different positions. The frequency response functions with respect to motor current are measured in ten different configurations.

5.2.1. Modal Analysis Setup

Three different robot positions were investigated. The positions were selected as to have important shifts in dynamics between each.

The first position is a fully stretched out position (Figure 4.a), which was estimated to have maximum inertia around a vertical axis. In the second position the arm is kept horizontally, but retracted, such that motor 2 is almost at its rear limit (Figure 4.b). This position was estimated to have a much lower inertia around the vertical axis.

The last choice aimed at increasing the inertia around a horizontal axis, while maintaining the inertia around the vertical axis small. Therefore the arm was raised (Figure 4.c). Maximum elevation was limited by the ceiling height of the test location.

The robot was bolted to the floor, in such a way that its location corresponded to a crossing of two main concrete beams, carrying the test floor. Nevertheless, as the results will show, the test floor can not be regarded as infinitely stiff. The boundary condition therefore must be regarded as elastic.

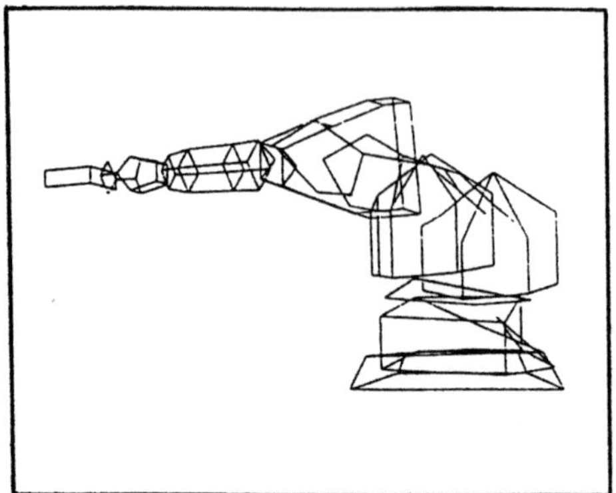

Figure 4a : IR 161, Configuration 1

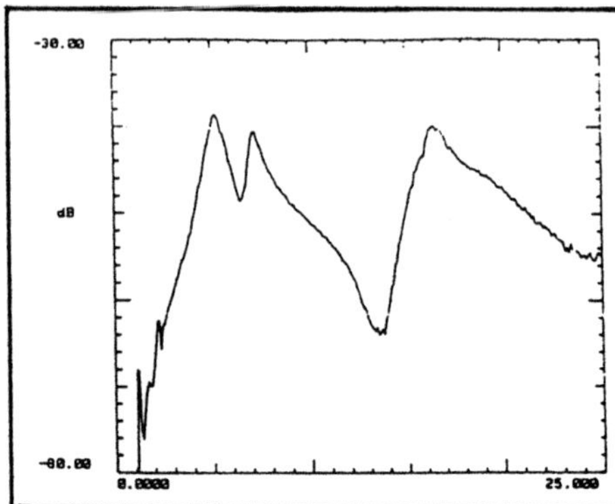

Figure 5a : Direct FRF, Configuration 1

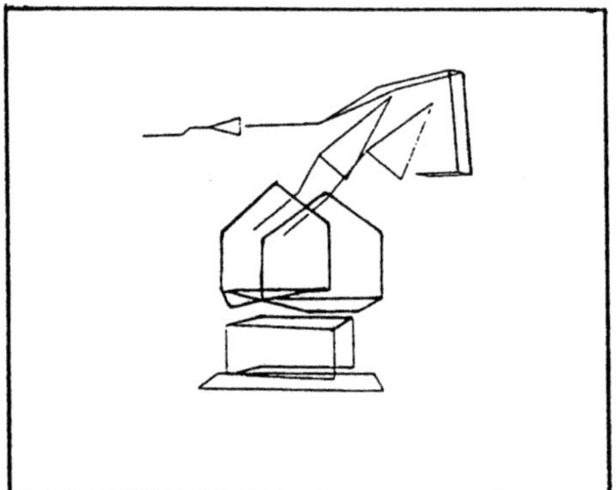

Figure 4b : IR161, Configuration 2

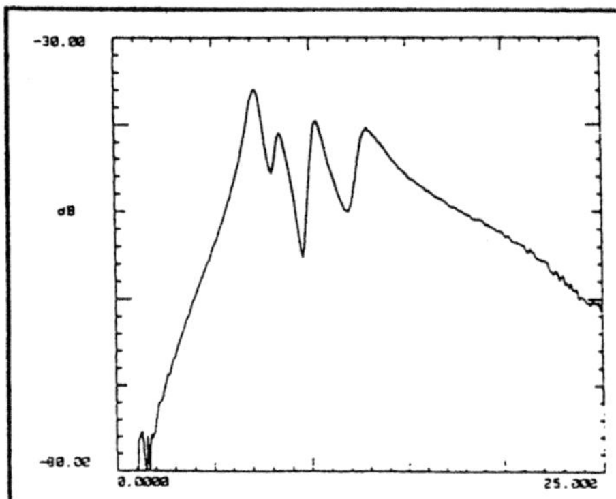

Figure 5b : Direct FRF, Configuration 2

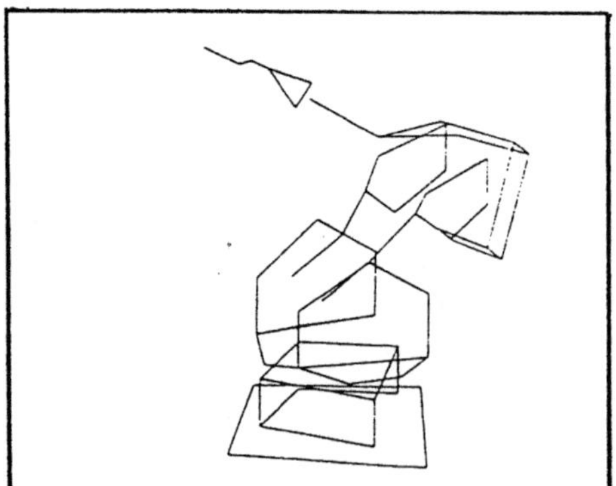

Figure 4c : IR 161, Configuration 3

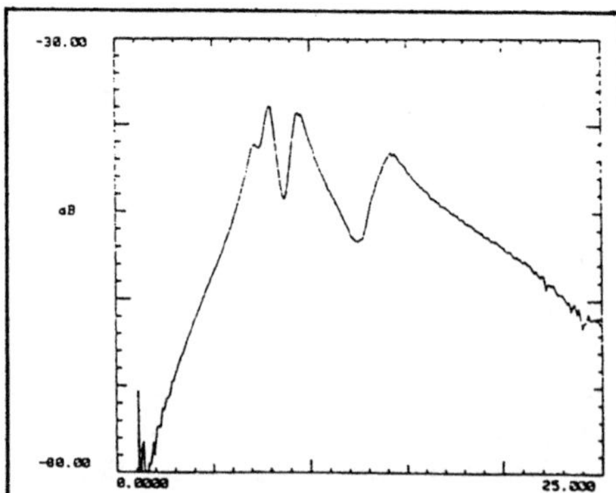

Figure 5c : Direct FRF,Configuration 3

5.2.2. Modal Analysis Measurements

Modal analysis was carried out using single input excitation.

Several excitation locations were compared prior to each configuration. This was done by using hammer impact in different places while measuring 7 typical responses in three directions. This very fast way of measurement allows to select a location which excites well all of the modes.

Responses are measured in several points in three directions by means of high sensitivity accelerometers *(300 mV/g, 1 to 5000 Hz)*. A set of 21 accelerometers was used to acquire simultaneously data for 7 points in three directions.

A periodic random excitation signal with a uniform amplitude and random phase spectrum was used. The actual injected force is measured by a force cell between the exciter and the structure. The frequency range of actual interest is limited to *25 Hz*.

In the first configuration 135 response points in three directions, resulting in 405 FRF's were measured. The geometry is depicted in Figure 4.a.

In configuration 2 and 3 a subset of the preceding points was taken. Since on site preliminary analysis proved that the modes did not affect the dimensions of the cross sections, it was possible to limit the total number of points.

5.2.3 - Modal Analysis Results

In each of the configurations five natural frequencies are found between *0* and *25 Hz*. Table 1 gives an overview of the natural frequencies and damping values for each of the configurations. The corresponding modeshapes are depicted in Figure 6.

Mode	Configuration 1	Configuration 2	Configuration 3
1	*5.17 Hz - 7.9%*	*7.19 Hz - 4.5%*	*7.25 Hz - 3.8%*
2	*7.05 Hz - 3.9%*	*8.31 Hz - 3.8%*	*7.98 Hz - 3.5%*
3	*13.33 Hz - 4.0%*	*13.45 Hz - 4.6%*	*13.12 Hz - 3.2%*
4	*16.32 Hz - 4.4%*	*10.20 Hz - 2.9%*	*9.32 Hz - 3.9%*
5	*17.84 Hz - 4.7%*	*12.65 Hz - 3.4%*	*13.99 Hz - 4.6%*

Table 1: Natural Frequencies and Damping

Figure 5 depicts the direct FRF's in the three measurement configurations. Except for the *13.3 Hz* mode, all modes are very well excited in each of the configurations *(x,y,z)*. It proves that some of the modes are strongly direction dependent.

The first mode of configuration 1 is a rotation around the vertical axis [Figure 6.a]. The main flexibility is located in the belt drive of motor 1. The same mode as mode one is found back in configuration 2 and 3. In the third configuration mode 1 is however slightly diagonal instead of purely horizontal. Since the inertia around the vertical axis decreases, frequency shifts up.

The second mode in configuration 1 is a flexion in the vertical plane [Figure 6.b]. The flexibility can however not be attributed to one single joint: all joints seem to be important. The same type of mode has also been observed in configuration 2 and 3, but the configuration 3 mode is slightly diagonal. In configuration 3, the diagonal direction of the first and second mode are orthogonal. Both modes also become close in frequency.

The third mode of configuration *1* at *13.33 Hz* is actually a flexion of the test floor, interacting with the robot dynamics. This test floor mode is observed at *13.45 Hz* and *13.12 Hz* in configuration 2 and 3 respectively.

Mode 4, configuration 1 at *16.32 Hz* is a horizontal flexion [Figure 6.c]. It shifts to *10.20* and *9.32 Hz* in configurations 2 and 3. Note that while the first mode was a pure rotation in a horizontal plane, this mode is an actual bending in the horizontal plane.

The last mode is found at *17.84 Hz* in the first configuration [Figure 6.d]. It represents a bending in the vertical plane. It is observed at *12.65 Hz* and *13.99 Hz* in configuration 2 and 3.

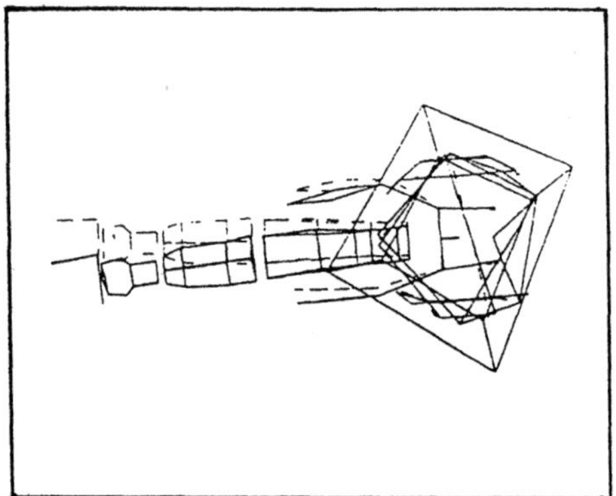

Figure 6a : IR161, Mode 1 (top view)

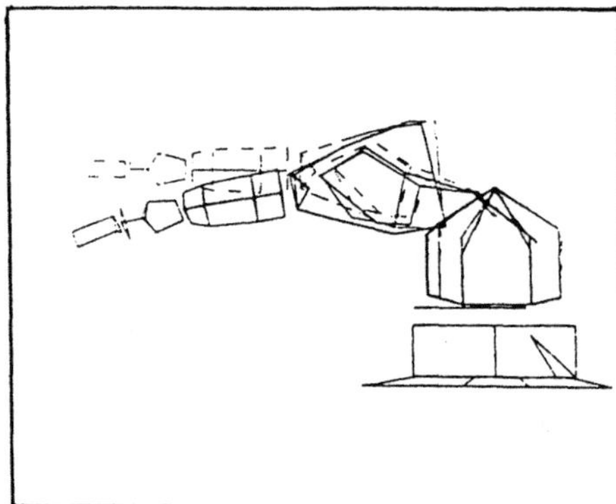

Figure 6b : IR161, Mode 2 (side view)

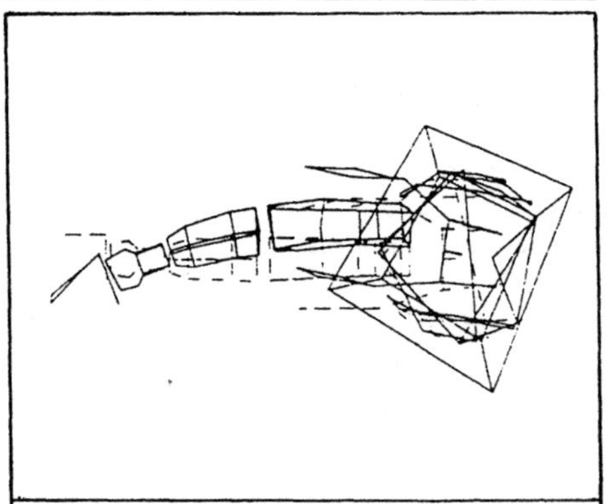

Figure 6c : IR161, Mode 4 (top view)

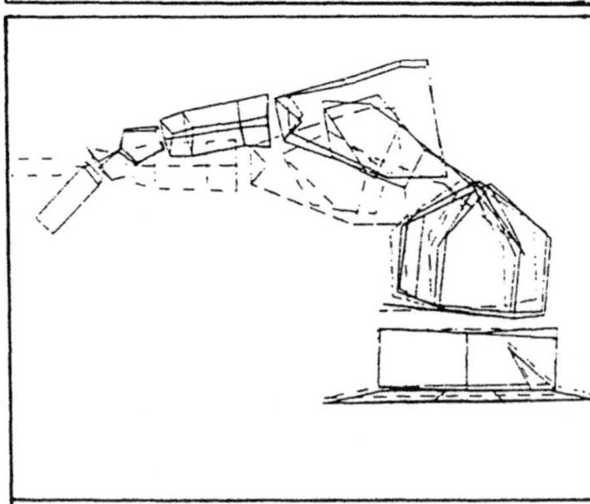

Figure 6d : IR161, Mode 5 (side view)

Globally, in configuration 1 the modes are spread out over a larger frequency range, as compared to the other configurations. In configuration 2 and 3, similar modes are observed, but much closer to each other: the higher frequency modes come down in frequency and the lower frequency modes shift to higher frequencies.

5.2.4. DYNAMIC CHARACTERISATION BY MOTOR EXCITATION

The modal model of the previous paragraph describes the dynamics in terms of displacement per unit externally applied force. It is, however, important for control purposes to have displacement response per unit motor current available.

Therefore a series of FRF measurements with excitation through the robot motors was carried out. The first three motors were individually used for excitation. 10 Robot positions were investigated.

In summary, the higher frequency global horizontal and vertical bending modes can be observed in the acceleration over current FRF's. The first two modes, which are built up by local flexibilities in the joints are only peaks for excitation by motors which are not situated in the node of the mode. A more detailed discussion is presented in [11].

6. Nonlinear System Behaviour [12,13,14,15]

6.1. Overview

Since all traditional system models used to describe the dynamic behaviour of systems assume that the system under study is linear, the influence of eventual deviations from this assumption was considered. This is especially important for systems such as flexible robot structures because of the possible nonlinear nature of the drive system.

First, a review of this subject was made, covering both the documentation of case studies, describing identification problems due to nonlinearities, as well as a compilation of possible detection and characterisation methods [12].

One measurement technique, based on sinusoidal excitation and two analysis techniques, resp. the Hilbert transform and the complex stiffness method were selected for implementation on the computer aided test systems at K.U.Leuven and LMS. In addition, a new time domain analysis technique was investigated. These techniques have then been applied to the analysis of the nonlinear behaviour of several of the developed Breadboard Models as well as to the identification of the KUKA IR360 robot.

6.2. Methods

6.2.1. DATA MEASUREMENT

During the review it was concluded that experimental techniques using purely sinusoidal excitation are the best suited for analysing systems exhibiting nonlinear behaviour.

As a result of Esprit Project 322, CAD*Interfaces, a new sinusoidal technique, referred to as stepped sine testing and using incremental stationary sinusoidal excitation, was developed and integrated into the LMS CADA-X software package for computer aided testing. Consequently, this technique is applied in the investigation of the possibly nonlinear behaviour of flexible robot structures. The method was extended with special features to measure and store the full harmonic signal spectra as well as the complete time records for each measured frequency.

6.2.2. Data Visualization

For the special case of sinusoidal excitation, substantial information can be gained from a study of the evolution of the complete signal spectra as a function of the excitation frequency. With nonlinear systems, these signal spectra not only contain the excitation frequency component, but also a number of harmonics. This phenomenon corresponds to a redistribution of the signal energy from the excitation frequency to other frequencies.

In order to visualize these frequency-in to frequency-out effects in an appropriate way, a pseudo 3D, or so-called Waterfall display was provided, adapted from an existing software for rotating machinery analysis (CADA-TEST Signature Monitor). Using this display, interactive operations allow to retrieve various analysis functions, such as cuts along the excitation frequency axis, along the response frequency axis or even along a harmonic component. This way, the energy transfer and interaction between the input (excitation) frequencies and the output (response) frequencies can be investigated.

This graphical software is demonstrated in the examples in the subsequent sections. The same display can also be used to present multiple frequency response functions (e.g. with the excitation level as parameter) in a more appealing way.

6.2.3. Data Analysis

In addition, two analysis techniques were implemented. One technique consists of the Hilbert transformation, the other of the Complex Stiffness technique. Both techniques are integrated as application programs within the open programming environment offered by the "User Programmable Mode" of operation of the LMS CADA-X Fourier Monitor software package. In addition, a new, direct time domain identification technique was studied at the K.U.Leuven and a prototype software was developed.

The theory of the Hilbert transform and the Complex Stiffness technique is discussed in [12,13]. Basically, the Hilbert transform of the frequency response function (FRF) of a linear and causal system should coincide with the FRF itself. A difference between them can be evaluated as a characteristic of nonlinear behaviour. The further interpretation into the nature of the nonlinearity is however not so straightforward.

The Complex Stiffness technique [13,14] on the other hand maps the system input/output frequency spectra (and FRF) into a presentation expressing the evolution of an equivalent stiffness and an equivalent damping as a function of resp. the input quantity and its derivative (e.g. displacement and velocity).

Further research is also performed to derive the necessary nonlinearity information directly from time history data. The theoretical background of the developed Time Domain technique can be found in [14].

6.3. Application to Breadboard Models

6.3.1. Introduction

Several breadboard systems have been analysed. In all of them, an important nonlinear system behaviour was encountered. With increasing excitation level, the FRF between the motor torque and the motion responses on the beam, changes in appearance occured with a resonance shift to lower frequencies and with a decreasing damping.

Below, a typical case is discussed. The system studied is the two-link breadboard model (BBM-4). Since a Coulomb friction in the direct drive system was detected to be the source of an important nonlinear behaviour, an actual experimental identification and modelling was performed and verified by means of simulations. Two different sets of measurements were performed: burst random and stepped sine tests.

6.3.2. Burst Random Tests

Two frequency response functions (FRF's) were measured: one FRF between the motor drive input signal (torque) and an LVDT displacement transducer on the first link, and one FRF between the torque and an acceleration transducer on the first link. The latter, together with its coherence, is shown in Figure 7.a. The quality of the FRF's is poor and the coherences are too low to be explained by measurement noise. This already gives an indication of the existence of nonlinear behaviour.

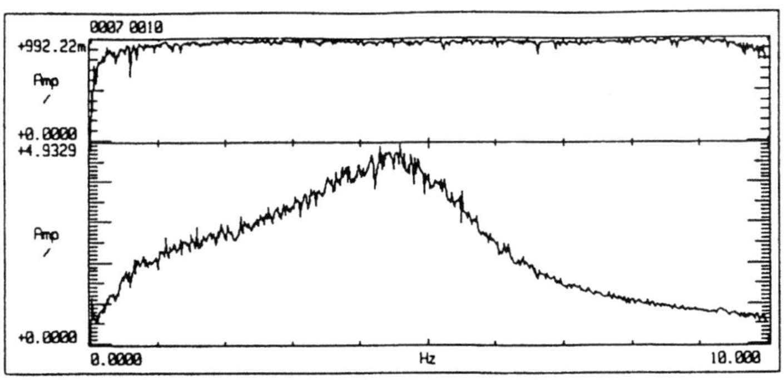

Figure 7a : Two-Link Breadboard : Broadband FRF

6.3.3. Stepped Sine Tests

Similar measurements were performed using stepped sine excitation at various levels and in two positions. The accelerometer FRF's corresponding to an extended position of the second link (max. inertia) are shown in Figure 7.b. The excitation levels, expressed in motor input voltage, were *0.2 V*, *0.3 V*, *0.35 V* and *0.4 V*.

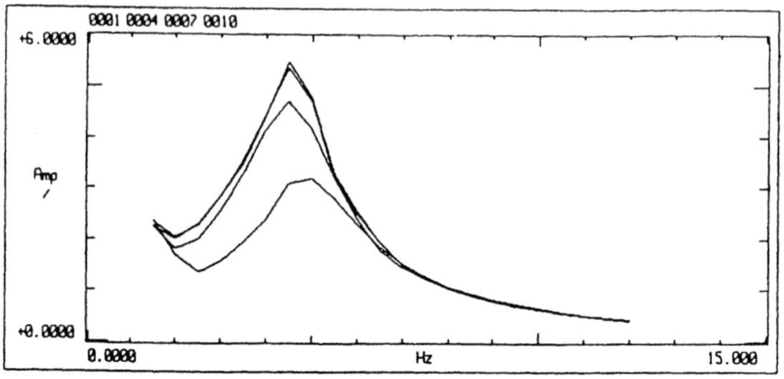

Figure 7b : Two-Link Breadboard : Stepped Sine FRF

Also in this breadboard, an decrease in maximum level, and an increase in resonance frequency and damping is observed with a decreasing excitation level, confirming the conclusions of the earlier tests that a stick-slip phenomenon in the motor drive system is the basis of the nonlinear behaviour.

However, while explaining the nonlinear damping behaviour, this does not directly explain the direction of the resonance shift, which cannot be due to a nonlinear stiffness phenomenon.

In order to model this nonlinear behaviour a series of additional specific tests and analyses were performed; the resulting models were verified by simulations, as described in the subsequent sections.

6.3.4. Friction Modelling: Measurements

An estimate for the Coulomb friction in the direct drive motor DYN ASERV DM A1200, to include into the physical model of the 2-link system, is obtained by the following procedure.

Stepped sine measurements are carried out at different input levels, frequency response functions are measured and direct time history data are recorded. The input to the motor drive is sinusoidal (torque input), the output measured is the tangential acceleration.

In this particular measurement set-up two different configurations are looked at. In the first configuration, the flexible arm, connected to the motor, is hinged to the ground. On this system both the frequency domain and the time domain identification techniques are applied. In the second configuration, the motor drive is tested with the flexible arm detached. In this configuration only time domain measurements are carried out (no resonance peak can be expected). The results of the methods are compared to direct free decay measurements on the motor drive.

Figure 8 gives the frequency response functions for the motor -arm configuration at three different input levels *(3.2, 4.8, 6.4 Nm)*. Figure 9.a gives the extracted stiffness characteristics and Figure 9.b shows the damping torques, using the complex stiffness method. The results are similar to those obtained when characterising the other breadboard configurations.

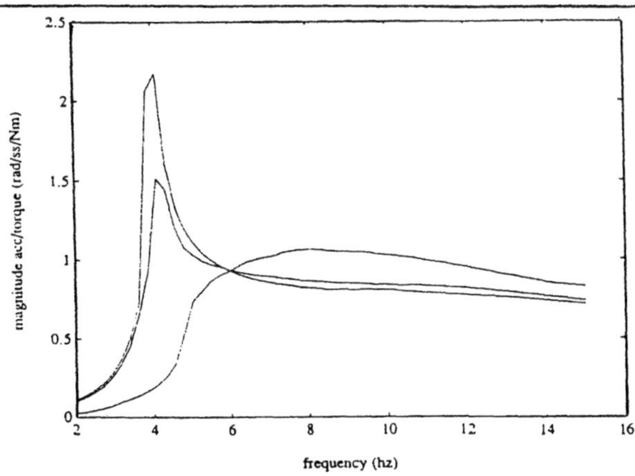

Figure 8 : Friction Modelling : Stepped Sine FRF

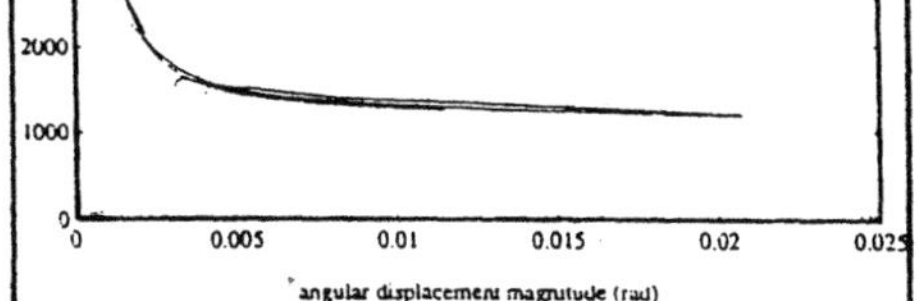

Figure 9a : Equivalent Stiffness (CS)

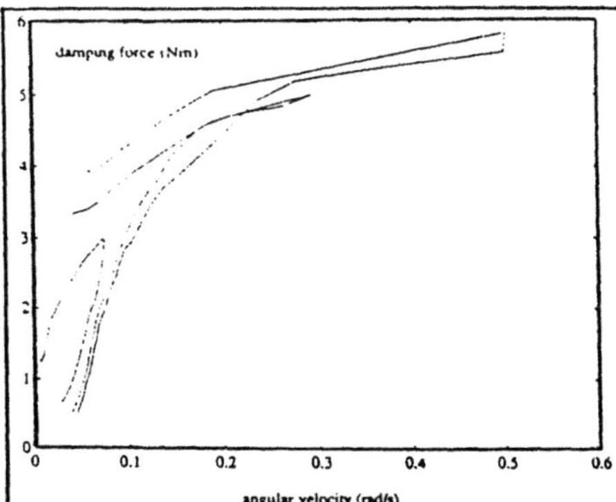

Figure 9b : Damping Torque (CS)

Figure 10.a shows the extracted restoring torques, applying the time domain identification method for the input level *3.2 Nm*. For the very low input levels the slopes of the restoring torques vary as a function of the input frequency, a phenomenon similar to the results on the previous breadboard. This stiffness variation is related to the fact whether or not sticking occurs (system velocity remains zero, until the external torque can compensate for the friction torque). This stiffness variation could be possibly explained by the existence of a spring in series with the dry friction damper. This extra spring then comes into play when stick occurs. Figure 10.b shows the damping torques, where again Coulomb friction prevails.

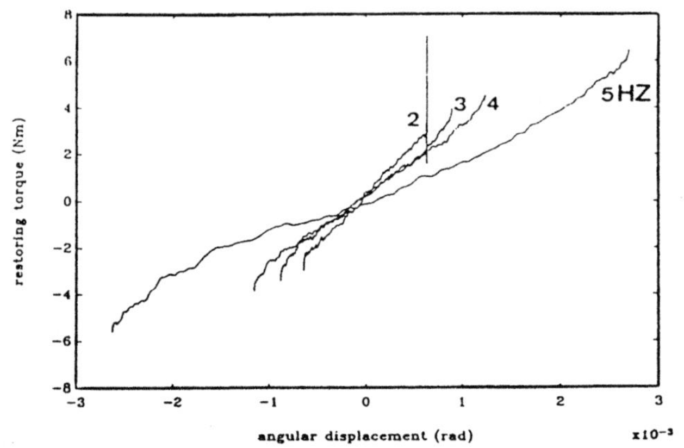

Figure 10a : Hinged Beam Restoring Torque (DT)

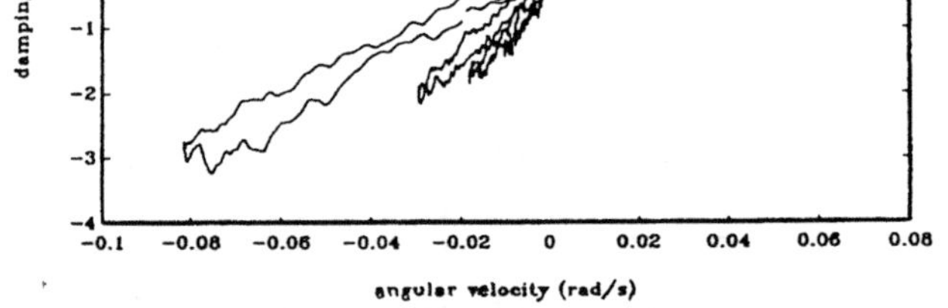

Figure 10b : Hinged Beam Damping Torque (DT)

From time history measurements on the motor drive only, the characteristics derived can be seen in Figures 10.c and 10.d. From Figure 10.c it is clear that the torque ripple due to the electromagnetic effects of the poles results in a stiffness variation. The damping torque characteristics suggest again that the Coulomb dynamic friction equals $5\,Nm$. When comparing this to the characteristic obtained when measuring the free decay response of the motor (given an initial velocity, the external torque is put to zero,

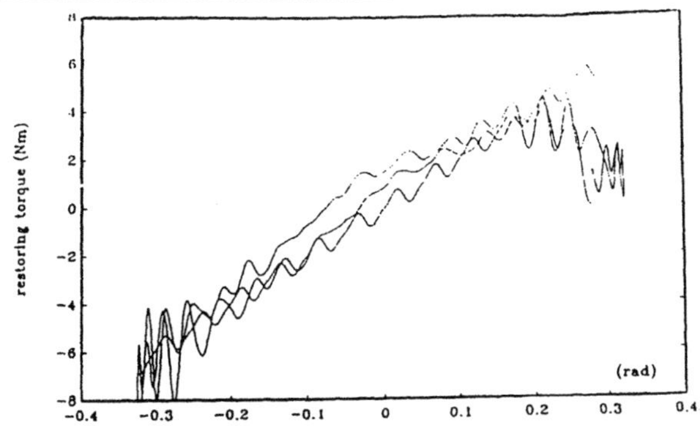

Figure 10c : Motordrive Restoring Torque (DT)

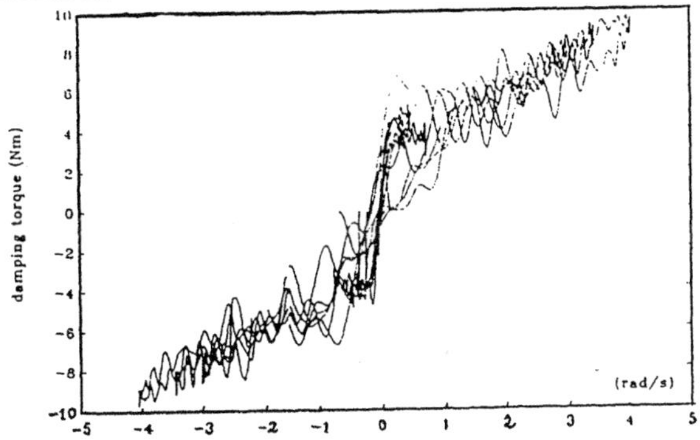

Figure 10d : Motordrive Damping Torque (DT)

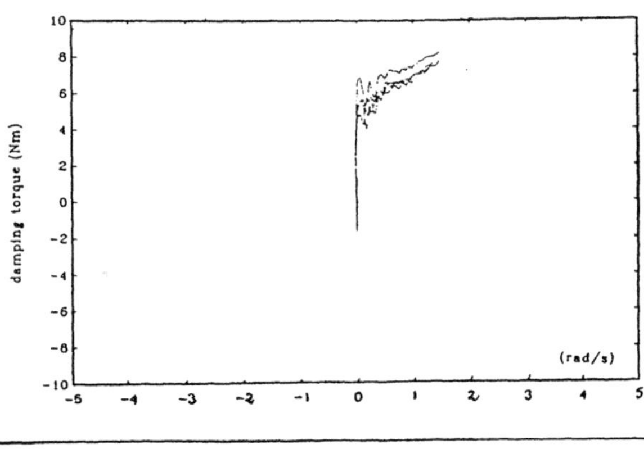

Figure 10e : Free Decay Damping Torque.

and the free decay acceleration is measured) (inertial torques equal damping torques), very good correspondence is obtained (Figure 10.e).

6.3.5. Friction Modelling: Simulations

Time domain simulations of the SDOF system with Coulomb friction included are carried out using the following equation of motion. The varying slope of the restoring torques is not taken into account.

$$J\ddot{T}+t_k+c\dot{T}+t_{dyn}sgn(T) = t_{ex}\,sin(\omega t)$$

with T the angular motion ($\dot{T}$ and $\ddot{T}$ resp. first and second derivative), J the inertia, t_k the restoring torque, c the viscous damping, t_{dyn} the dynamic friction.

When looking at the synthesized frequency response functions at different input levels (Figure 11.a), a discrepancy between measured and simulated results comes forward. The measurements are characterised by a stiffening effect with decreasing excitation input level, while the simulations show a reverse softening effect with decreasing input level.

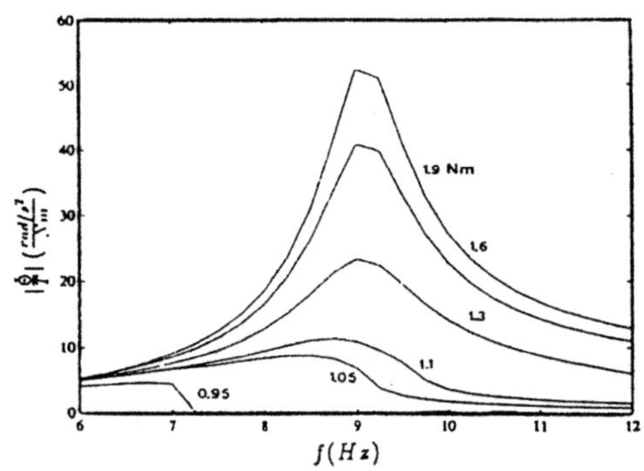

Figure 11a :Dry Friction Model : FRF

In order to try to explain this contradictory effect, the model of Figure 11.b is proposed. For this particular simulation case, a harmonic probing method is used to extract the frequency response functions (sinusoidal displacement) at different inputtorque levels. Figure 11.c shows the simulation results for an extra stiffness k_{ex}. In this case, the extra spring comes into play at the moment when sticking occurs, and the "resonance" frequency increases with decreasing input levels.

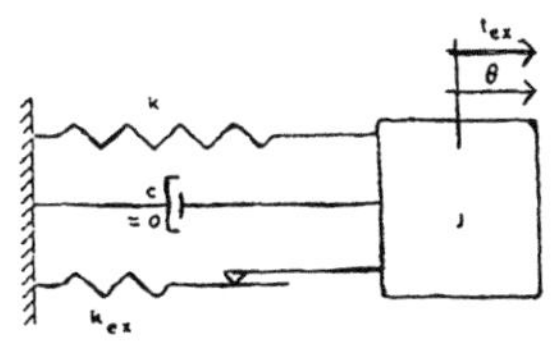

Figure 11b : Modified Model

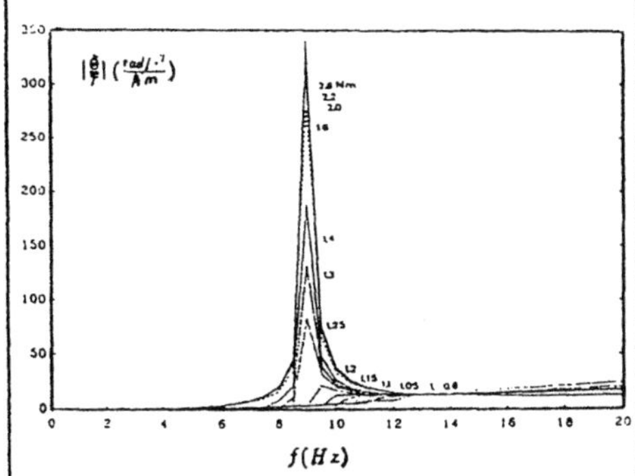

Figure 11c : Modified Model : FRF

6.3.6. CONCLUSIONS

The identification methods proposed prove again to be powerful tools for the identification of nonlinear mechanical systems, when they can be approximated by a discretised SDOF model. The results obtained from measurements on both direct drive systems are very similar and reveal the same tendencies: the dry friction is predominant in both systems, the electromagnetic torque ripple is apparent in both drives (124 poles), and both systems show a stiffness rise for very low input levels, related to sticking. An explanation for the discrepancies between simulation (apparent stiffness drop when sticking occurs) and measurements (apparent stiffness rise when sticking occurs) of dry friction systems is proposed.

6.4. Application to the KUKA - IR360 Robot

6.4.1. DESCRIPTION OF THE SET-UP

The purpose of the additional tests was to study the dynamic behaviour of the complete system: controller - motor - transmission - robotstructure. In view of the conclusions of the previous robot modal analysis tests, identifying the flexibilities of the first modes in the first axis drive system, excitation was applied through the controller and motor of axis 1. Response signal measurements were taken at the end of the robot arm. At this location, the acceleration in two directions was measured by means of two high performance accelerometers. In addition, acceleration responses were measured at two intermediate points: one in the tangential direction of the large reduction wheel of axis 1, at the bottom of the robot, and one at an excentric location on the shaft of axis 1, above the first joint.

6.4.2. Broadband Measurements

At first, broadband excitation frequency response measurements were performed using a Hv FRF estimator. They revealed a particularly low coherence value, *(0.8 to 0.9)*, which could not be explained by signal processing (leakage) errors, or by environmental noise errors. In Figure 12.a, the FRF and coherence of the transmission path between controller input and vertical end-point acceleration is shown.

Measurements of the signal levels in the absense of a forced excitation (no controller input) were used to establish the environmental noise level, which proved to be several orders below the signal levels and thus not the cause of the low coherence.

Repetition of the measurements in identical conditions yielded nearly identical results, while repetition with different excitation levels revealed a strong input level dependency of the FRF.

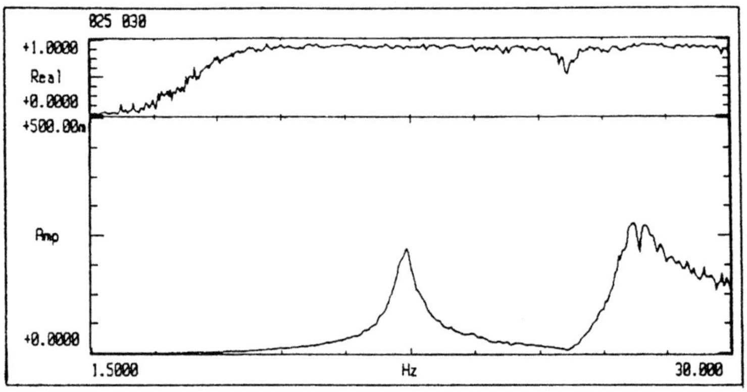

Figure 12a : Broadband FRF (bottom) and coherence (top).

6.4.3. Stepped Sine FRF Measurements

In order to study these apparent nonlinear effects in more detail, a series of stepped sine excitation tests was performed. The time histories of the measured responses already show interesting information. In Figure 12.b, the time histories (from the top downward) of the acceleration at the reduction wheel, at the shaft, at the vertical end-point and at the horizontal end-point are shown for a pure sine excitation of *3 Hz*. The "noisy" outlook can clearly be attributed to the large amount of harmonic distortion. Especially in the third (vertical end-point) signal, it can be seen that this harmonic component which excites the first resonance *(15 Hz)* has the most important contribution.

The stepped sine measurements were then repeated for three different excitation levels. In Figure 12.c, the three different stepped sine FRFs are compared to these obtained with broadband testing for the vertical end-point acceleration. For the other signals, even larger differences were found.

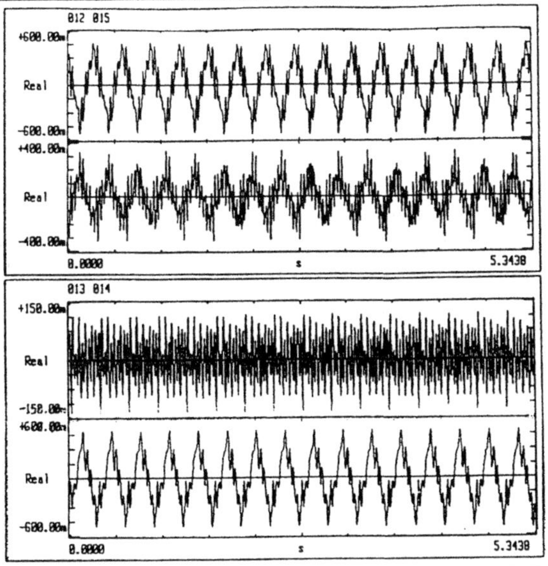

Figure 12b : Time Signals, 3 Hz Excitation.

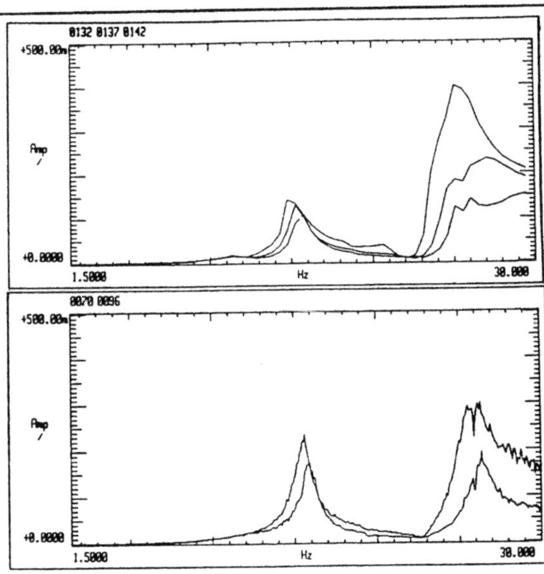

Figure 12c : Stepped Sine FRFs at 4V, 3V and 2V Excitation (top) compared to Burst Random FRFs at 3V and 5V (bottom).

6.4.4. Stepped Sine Harmonic Analysis Measurements

Finally, an additional series of sine tests was performed to specifically study the generation of harmonic sine components. Basically such a harmonic spectrum analysis consists of a stepped sine test procedure, scanning the frequency range of interest, and storing the full signal spectrum at each excitation frequency.

64 Data samples are taken per sine period allowing correct (alias free) estimates up to the thirtieth harmonic, 16 sine periods are observed each time, allowing an inter-harmonic separation of 16 (up to the sixteenth subharmonic can be resolved).

The corresponding results are shown in Figure 13 for the case of the vertical end-point acceleration. The evolution of a large number of harmonics can clearly be distinguished. When a harmonic component coincides with a structural resonance, it is drastically magnified. This is in agreement with the timeseries analysis.

6.4.5. Frequency Projection Analysis

More information can be gained from the input/output frequency analysis. This can be done by making sections in the 3-D harmonics map, along the axis of the various harmonics components. These sections can then be projected on either the input frequency axis or the output frequency axis. The former option allows to investigate which frequencies are excited, for each harmonic separately, by each possible input frequency, while the latter option allows to analyse which harmonics components are responsible for the response at one specific frequency.

In Figure 14.a, the 3-D harmonic map of a detailed test in the low-frequency region is shown (vertical end-point acceleration), more clearly indicating the resonances excited by the higher harmonics. The harmonic cut projections along the input frequency axis *(1.5-30 Hz)*, resp. output frequency axis *(0-180 Hz)* are shown in Figures 14.b and 14.c.

In Figure 14.b, the same resonance phenomena are found back at fractional values of the basic resonance frequency. This means that harmonic components, excited by the nonlinearity, excite the structural resonance when their frequency is close to it.

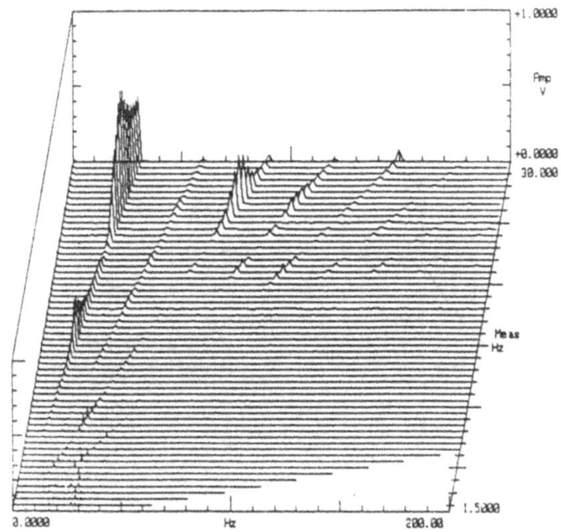

Figure 13 : Vertical End Point Acceleration; Harmonic Analysis.

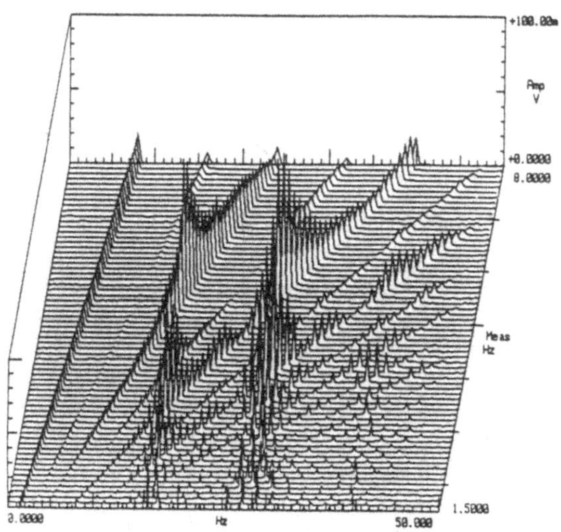

Figure 14a : Harmonic Analysis, 3D Spectral Map.

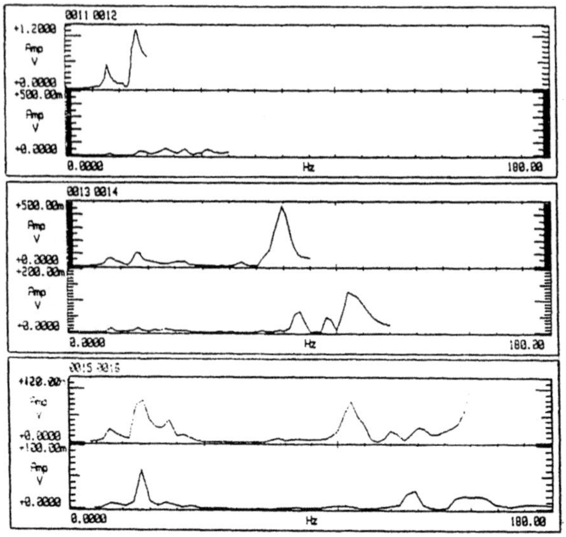

Figure 14b : Harmonic Analysis, Input Frequency Projection (Harmonics 1-6).

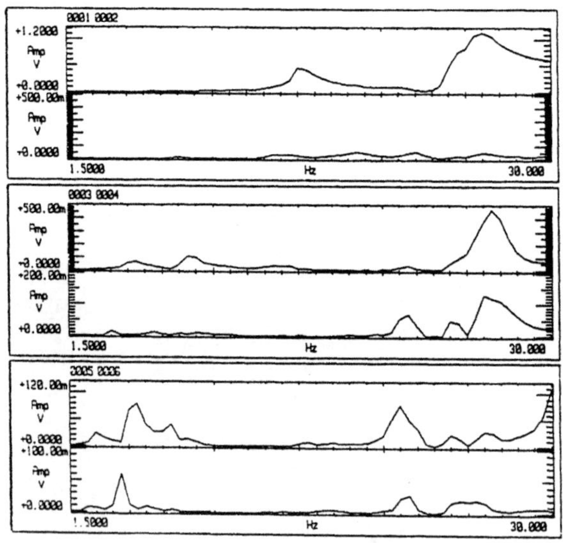

Figure 14c : Harmonic Analysis, Output Frequency Projection (Harmonics 1-6).

When analysing these data for peak values (resonances), the following values result:

Harmonic	Vertical Acceleration
1	*15.0 Hz - 26.5 Hz*
2	*5.0 Hz - 8.5 Hz*
3	*3.0 Hz - 5.5 Hz*
4	*2.5 Hz - 4.0 Hz*

For example, a *5 Hz* sinusoidal input will result in an important vertical movement of the robot at *15 Hz* because the third harmonic *(15 Hz)*, generated by the nonlinearity, excites the first vertical mode.

This non-parametric way of analysing the interaction between harmonic distortion and structural resonance could be a first step in a higher-order model formulation to describe nonlinear phenomena.

6.4.6. Conclusion

Both broadband FRF and stepped sine FRF measurements reveal an important nonlinear behaviour of the complete system: controller - motor - transmission - robot structure.

Based on the conclusion of the structural dynamic tests, it can be concluded that the origin of the nonlinearity is most probably not structural. The tests on breadboards on the other hand show important nonlinear behaviour of the drive (motor) system. Hence, also for the robot, the nonlinearities are expected to be generated in the motor subsystem.

Final conclusions as to the nature of the nonlinearity are however not yet reached and additional tests and analyses will be performed. It can however clearly be seen that this matter is of the highest importance if a proper robot control is to be achieved.

7. Exploitation of Results

Besides of the evident application of the outcome of the identification results in the simulation and control tasks of the project, the developed identification methodology will be exploited as follows:

- Through an increased know-how in the areas of the experimental identification and numerical modelling and simulation of flexible multibody systems and combined control/structural systems. This includes an increased experience in this specific class of applications, new and improved measurement procedures and technology, techniques for evaluating and analysing the nonlinear system behaviour as well as the application of identification results to multibody dynamic systems.
- Through an increased functionality and performance of the related Computer Aided Dynamic Analysis software by the integration of the Frequency Domain

Direct Parameter Identification technique. However, also other functionality improvements have been made, for example in view of the analysis of nonlinear behaviour (extension of stepped sine testing software for level dependency and harmonic distortion analysis, pseudo-3D harmonic data presentation, specific data analysis procedures) and in view of the application of simulation results in multibody dynamic simulations (addition of finite element data preprocessor for neutral file to generate the required kinetic moments).

8. Conclusions

In order to derive the necessary models for the dynamic characteristics of flexible robot structures, advances have been made in the development of identification techniques. Two methods, based on a state space model formulation, were developed and implemented, one method making use of time domain and the other making use of frequency domain data.

The developed methodology was applied to the identification of the dynamics of the studied breadboard systems as well as of the selected robots.

Special attention was paid to the problem of nonlinear system behaviour. Detection and analysis techniques were evaluated, implemented and tested on the breadboards and the robot structures.

An interface between the dynamic modelling and the simulation software was realised, using the concept of the CAD*I Neutral File.

Bibliography

[1] Leuridan J., Van der Auweraer H., Mergay M., "Review of Parameter Identification Techniques", Technical Report of Esprit Project 1561, SACODY, Ref.6111, Lecture Notes of the 15-th Int. Seminar on Modal Analysis, Basic Course, Vol.2, Sept.17-18, 1990, ed. Div. PMA, K.U.Leuven, Belgium

[2] Lembregts F., Brughmans M., "Estimation of Real Modes from FRFs via Direct Parameter Identification", Proc. 7-th IMAC, Las Vegas (ND), Feb. 1989, pp.631-637, ed. Union College, Schenectady (NY).

[3] Lembregts F., Leuridan J., Van Brussel H., "Frequency Domain Direct Parameter Identification for Modal Analysis: State Space Formulation", Mechanical Systems and Signal Processing, Vol.4, No.1, pp.65-75, 1990.

[4] Lembregts F., Mergeay M., Van der Auweraer H., Swevers J., Adams M., Sas P., "Development and Evaluation of Off-Line Identification Techniques", Technical Report of Esprit Project 1561, SACODY, Ref. 612, LMS International, Belgium.

[5] Swevers J., "An Introduction to System Identification and its Applications in Practice", Lecture Notes of the 1990 Integrated European Course in Mechatronics, pp.41-70, ed. Div. PMA, K.U.Leuven, Belgium.

[6] Adams M., Swevers J., De Schutter J., Van Brussel H., "Identification and Control of a Single Degree of Freedom Positioning System", Proc. 13-th Int. Seminar on Modal Analysis, Sept.19-23, 1988, ed. Div. PMA, K.U.Leuven, Belgium.

[7] Brughmans M., Lembregts F., "Using Experimental Normal Modes for Analytical Model Updating", Proc. 8-th IMAC, Orlando (FL), Jan.1990, pp.389-396, ed. Union College, Schenectady (NY).

[8] Brughmans M., Leuridan J., "Development and Test of Interface Between CAT/CAE and System Modeling Simulation", Technical Report of Esprit Project 1561, SACODY, Ref. 5471, Sept. 1990, LMS International, Belgium.

[9] Leuridan J., Brughmans M., Bakkers W., De Landsheer A., "FEM Model Correlation and Updating in an Heterogeneous Hardware and Software Environment", Proc. 14-th Int. Seminar on Modal Analysis and Numerical Methods in Structural Dynamics, Sept.11-15, 1989, ed. Div. PMA, K.U.Leuven, Belgium.

[10] Van der Auweraer H., Vandeurzen U., Maes J., "Identification of Structural Dynamics Characteristics of Available Robot", Technical Report of Esprit Project 1561, SACODY, Ref.4811, LMS International, Belgium.

[11] Vandenhout J., Devis D., Gielen L., "Test and Off-Line Identification of Demonstration Robot", Technical Report of Esprit Project 1561, SACODY, Ref. 4851, Apr. 1990, LMS International, Belgium.

[12] Van Brussel H., Mertens M., Van der Auweraer H., Leuridan J., "Study of Available Techniques to Identify and Characterise Nonlinearities by Test", Technical Report of Esprit Project 1561, SACODY, Ref.6141, Lecture Notes of the 15-th Int. Seminar on Modal Analysis, Basic Course, Vol.2, Sept.17-18, 1990, ed. Div. PMA, K.U.Leuven, Belgium

[13] Mertens M., Van der Auweraer H., Vanherck P., Sas P., "The Complex Stiffness Method to Detect and Identify Nonlinear dynamic behaviour of SDOF Systems", Mechanical Systems and Signal Processing, Vol.3, No.1, pp.37-54, 1989.

[14] Vanherck P., Wyckaert K., Sas P., Van Brussel H., "Parametric Identification of Nonlinear Mechanical systems from Measurements in Time and Frequency Domain", Proc. 7-th IMAC, Las Vegas (ND), Feb. 1989, pp.695-703, ed. Union College, Schenectady (NY).

[15] Wyckaert K., Vanherck P., Sas P., Van Brussel H., "The Identification fo the Nonlinear Dynamical Behaviour of a flexible Robot Link", Proc. 8-th IMAC, Orlando (FL), Jan.1990, pp.530-539, ed. Union College, Schenectady (NY).

Chapter 4

Dynamic Simulation of Multibody Flexible Servomechanisms

Gérard Gallay, Jean-Luc Faillot
Bertin et Cie

1. Introduction

The generic term multibody systems encompasses a broad set of mechanical systems, ranging from robots, cranes, chains, cables, panels, antennae to human body models and other biosystems. The different kinds of multibody systems that can be distinguished are systems made of connected or separated bodies, systems made of rigid or flexible bodies and systems presenting closed-loop or open-tree topologies.

The design of multibody systems is more and more supported by computer dynamic simulations that allow the validation at an early stage of design trade-offs without implementing expensive test prototypes, that are moreover poorly representative of the target systems.

Such a need for a simulation tool was indeed stressed in the basic objectives of the SACODY project, who addressed the problem of high performance lightweight robots. Solving such a problem in a generic way actually supposed the availability of a software capable of:

- generality in terms of mechanical simulation capabilities,
- modelling multibody mechanical systems incorporating non rigid components,
- providing the background data for analysing such a mechanism or for correlating a model with experimental tests,
- yielding the data necessary for designing the system control,
- simulating the system dynamic behaviour in response to external forces (open-loop simulations) as well as under the effect of different control strategies (closed-loop simulations).

During the late 1970s and early 1980s, a lot of development had been devoted to the dynamic simulation of flexible multibody mechanisms, with application to spacecraft, satellites and manipulators. The earliest research in the field was carried out in the early 1970s for modelling the dynamics of spacecraft with flexible appendages. Although they addressed the case of small flexible motions around a given rigid configuration, these studies did not allow the analysis of large changes in the multibody systems configuration, which is one of the main characteristics of manipulators motions.

The treatment of the specific problems of the dynamic analysis of flexible robot-like mechanisms is more recent and dates back to the early 1980s. Most of the research in the field has been carried out in the United States, as further spin-offs of large space programs. Among the resulting software, only a few commercial packages allow for the dynamic analysis of truely flexible multibody systems, and none enables the design of control for flexible systems and its verification by closed-loop simulations.

These considerations have led to the enhancement within the framework of SACODY of a general computer code developed at BERTIN for the dynamic analysis of multibody systems, called ADAMEUS, which has been adapted according to the objectives set above.

The section presents the overall software, its background and functionalities, and describes its application for modelling the project demonstator as well as for designing its antivibration controller.

2. Overall Presentation of the ADAMEUS Software

2.1. Software Formulation

Since its initiation in 1983, the development of the ADAMEUS software has been focusing on the flexible multibody systems. The first objective was the development of a software tool for simulating the dynamic behaviour of such systems without, in this first step addressing the difficult problem of their control. The basic features of ADAMEUS were then as described in the following subsections.

2.1.1. Mechanical Simulation Capabilities

In terms of mechanical simulation capabilities, the largest level of generality was sought for ADAMEUS, the only limitation brought concerning the elastic character of the small deformations generated in some of the systems components during the motion.

As a consequence, ADAMEUS handles the most general multibody systems, that are the systems including flexible bodies, with closed mechanical loops and with translational or rotational degrees of freedom between their components.

2.1.2. Formalism

Most of the developments led in the field of computer simulation of flexible multibody systems have formulated the equations using either Newton's law, Lagrange equations or some modified geometrical theory. The difficulties usually encountered in the application of these formalisms are known to be:

- The introduction of non working forces between adjoining bodies,
- the need for equations derivation as in the Lagrange's formalism,
- the solution of the system equations in their developed form.

ADAMEUS has been developed using a formalism based on the virtual works principle of the Kane's method, that systematically avoids these problems.

This approach enables to get rid of the internal forces that act on non working degrees of freedom. Moreover, it provides much more flexibility for chosing the problem variables.

2.1.3. Characterization of Bodies

In terms of position, the system bodies, should they be flexible or not, are defined by a reference point and by a moving reference frame, linked to the body during its motion.

The reference point used in ADAMEUS is the body mass center, whereas other codes retain a material point, as for example one of the body connection points. The advantages of the choice made for ADAMEUS are twofold:

- First, it leads to a fair simplification of the equations of motions,
- then, it is often difficult to privilegiate a particular point of a body especially when systems with closed mechanical loops are to be addressed.

The moving reference frame will often be the frame in which the body inertia matrix is known and, in the case of a flexible body, the frame in which the body flexible behaviour is characterized.

The mechanisms bodies are dynamically characterized by their masses and inertia matrices expressed at the bodies mass center in the moving reference frames. Should a body be flexible, it will be further characterized by its modal parameters read by ADAMEUS in data files released by Finite Element Codes.

2.1.4. Choice of the Variables

The system variables used in the equations are relative linear or rotational velocities at the connections, for representing the rigid body motion, and the time derivatives of the body modal coordinates to represent the flexible body behaviour.

2.2. Software Organisation

2.2.1. Software Overview

ADAMEUS is a computer program whose purpose is to analyse the transient response of three-dimensional multibody mechanisms to internal or external forces and torques. Those mechanisms may have rigid or flexible elements and may have open-tree or closed-chain topologies. ADAMEUS implements, as the Finite Element Method, the principle of virtual works. This section describes the software organisation.

The execution of an ADAMEUS program encompasses two phases:

- The Preface, during which the program analyses and processes the mechanisms data and organises the data base.
- The Dynamic Analysis, principally performing the computation and the integration of the system motion differential equations.

2.2.2. PREFACE

The Data Input

The input of the problem data is divided into three parts:

The case control deck

The case control deck defines all the parameters for the case analysis. It defines the simulation to be performed and the solution adopted by the operator to carry out the analysis of the system in terms of:

- Nature of the simulation (With or without control),
- integration method (Runge-Kutta, Gear,...),
- integration step and simulation duration,
- listing and graphic outputs,
- etc...

The bulk data deck

The bulk data deck defines the mechanism to be analysed. It includes the definition of the bodies in their initial configuration (mass, inertia matrix, position of mass centers), the location of their connections (grids) and the type of joints, the external forces applied to the system, as well as the specification of the flexible bodies..

Interface with Finite Element Codes

For each of the flexible bodies identified by the program while processing the bulk data deck, ADAMEUS must have an access to the definition of the body modal deformation established by a Finite Element Code. To this aim, the interface program reads:

- The geometrical model of the body (i.e. the finite element mesh of the body),
- the generalized masses and stiffnesses of the retained modes,
- the natural frequencies and modeshapes of the body.

The data input ends with an echo of the model data.

The Analysis of the Mechanism Topology

Starting from the definition of the mechanism bodies and of their connection grids, this routine constructs the system topology, formalized by the body connection array (or lower body array) and detects the closed-loops presented by the system.

The Reduction of the Number of Degrees of Freedom

The total number of allowable degrees of freedom of a rigid mechanism with N bodies regrouped in N_s multibody subsystems is:

$$N_g = 3N + 3N_s.$$

In a preliminary step, ADAMEUS therefore builds N_g dynamic equations corresponding to the N_g G-level system degrees of freedom.

Nevertheless, the consideration of additional constraints such as one-dimensional rotation or translations at the connections or treatment of a purely planar problem leads to the definition of the *working degrees of freedom* labeled by ADAMEUS the N_f F-level degrees of freedom.

Constraints existing between the subsystems, which occur for instance in case of closed-loop or multipoint constraints, introduce an additional reduction of the systems degrees of freedom to the N_n degrees of freedom of the N-level, which correspond to the size of the dynamic problem to be solved.

2.2.3. The Dynamic Analysis

This module performs the dynamic simulation of the mechanism submitted to given effort conditions. In the basic version, the effort conditions allowed were open-loop constant or time-varying loads such as the gravity, external forces and moments applied on any point of the structure, etc... As it will be developed later in this chapter, these conditions now include the case of closed-loop forces dependent on the system degrees of freedom at each time step.

The results of the Dynamic Analysis are the positions, velocities and accelerations of the system rigid and modal variables as well as the linkage efforts at the connection points. They are obtained by the numerical integration of the system motion differential equations which can be performed by the Runge-Kutta, Gear and Adams-Moulton integration methods.

3. New Functionalities of the ADAMEUS Software

The previous chapter has focused on the initial features of the ADAMEUS software that mainly concerned the derivation for simulation purposes of the dynamic equations of non-rigid multibody systems submitted to external loads. During the SACODY project, new developments of ADAMEUS have been undertaken in order to provide this basic software with new capabilities paving the way for lightweight robots structure and control design. This section describes these new functionalities that are:

- The derivation of linear systems equivalent to the complete system around prescribed functioning configurations,
- the ability to incorporate software modules simulating feedback controllers servoing degrees of freedom of the system,
- the calculation of model-dependent control parameters necessary to design high-performance control schemes.

The overall organisation of the new software version coming out of SACODY is shown Figure 1.

3.1. Linear Models

The most outstanding characteristic of multibody systems is the high degree of nonlinearity of their dynamics: The mass matrices of these systems vary with their overall configuration and they are submitted to changes of the gravitational forces and highly nonlinear couplings (Coriolis and centrifugal forces) during their motion.

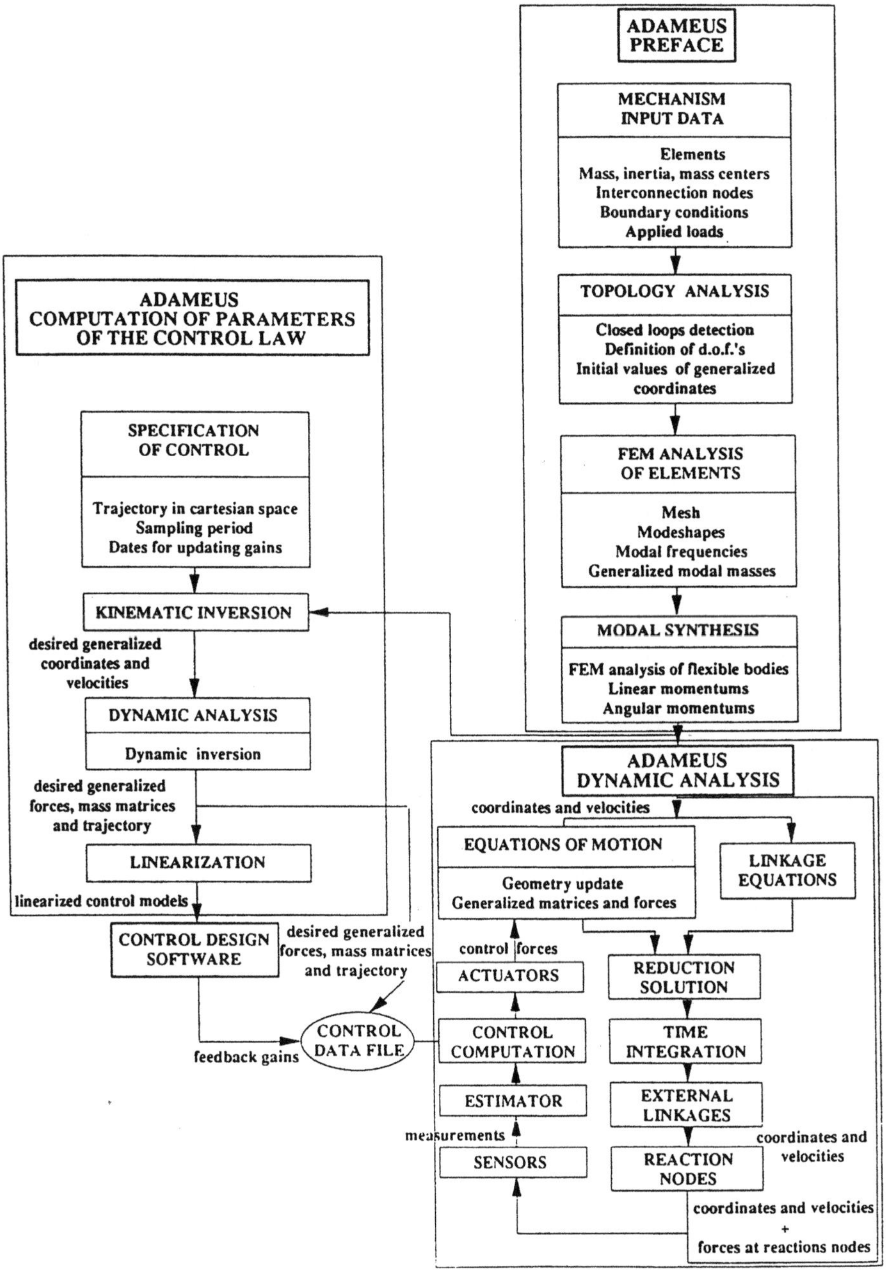

Figure 1: Overall Organisation of ADAMEUS Software

The nonlinear dynamic equations, numerically formulated in the ADAMEUS Dynamic Analysis, are therefore necessary to accurately reproduce the behaviour of the system.

On the other hand, system models based on such nonlinear equations are ill-adapted to the tasks of analysis for the structure and/or for the control design of such mechanisms.

This statement is perfectly illustrated by the needs that arised during SACODY when focusing on the particular case of lightweight robots. The requirements were then:

- To provide a means to analyse the global vibration modes of a model established after flexible bodies modes (or constrained modes) information,
- To enable the correlation and the tuning of the model parameters after linear models yielded by experimental Modal Analysis carried out on a few number of the system configurations,
- To provide linear state-space models representative of the system dynamic behaviour in the vicinity of prescribed functioning points, to be used to design the mechanism feedback controller.

The generic form of the dynamic equations of a multibody system established with bodies flexible modes may be written:

$$M(q)\ddot{q} + f(q,\dot{q}) + g(q) + d\dot{q} + kq = BT \qquad (1)$$

where:

- $q = [\theta^T \ \xi^T]^T$ is a vector regrouping the rigid degrees of freedom (θ vector) and the flexible bodies generalized coordinates (ξ vector),
- $M(q)$ is the system mass matrix,
- $f(q,\dot{q})$ is the vector of the Coriolis and centrifugal forces,
- $g(q)$ is the vector of gravitational forces,
- d is the system damping matrix,
- k is the system stiffness matrix,
- T is the vector of the torques applied at the joints.

Linearizing the system about a functioning point (q_0 , q_0 , T_0) consists therefore in computing the matrices M, K and D of the differentiated form of equation (1):

$$M(q_0)\delta\ddot{q} + D(q_0,\dot{q}_0)\delta\dot{q} + K(q_0,\dot{q}_0)\delta q = B\delta T \qquad (2)$$

with :

$$\begin{cases} D = d + \frac{\partial f}{\partial \dot{q}}(q_0,\dot{q}_0) \\ K = k + \frac{\partial f}{\partial q}(q_0,\dot{q}_0) + \frac{\partial g}{\partial q}(q_0) \end{cases}$$

This linearisation has been introduced in ADAMEUS along with its formulation in the augmented phases space, which is used for modal analysis purposes as well as for the feedback control design:

Writing $x = [\delta q^T \delta\dot{q}^T]^T$ leads to the matrix state-space equation:

$$\dot{x} = A\,x + B\,T$$

with $$A = \begin{bmatrix} 0 & I \\ -M^{-1}K & -M^{-1}D \end{bmatrix}$$

and $$B = \begin{bmatrix} 0 \\ M^{-1} \end{bmatrix}$$

The eigenvalues and eigenvectors of A provide respectively the modal pulsations and associated modeshapes of the system around x_0.

The matrix pair (A, B) may as for it be used for analysing and synthesizing feedbacks, state observers and miscellaneous filters by application of the Control Science linear methods (pole placements, LQR/LQG, LTR,...).

3.2. Integration of Feedback Control

As previously stated, the initial version of ADAMEUS allowed the simulation of multibody mechanisms behaviour in response to external loads, but not the simulation of feedback controlled systems, which means systems animated by control orders resulting from measurements made by sensors located on the moving bodies.

It was of course one of the basic objectives of SACODY to provide ADAMEUS with this new capability, that was required in order to offer to the lightweight mechanisms control designers a software tool for checking by simulation the effects of their control schemes on the system global behaviour.

The following developments have been undertaken in order to meet this requirement:

- Sensor modules have been introduced in the software. They enable the reproduction, using the system state variables, of sensor measurements such as position, velocity, acceleration, strain gages measurements etc... in any point of the system, including sensor gain, resolution and measurement noise.
- Actuator modules have been introduced in order to allow the simulation of the actuator dynamics and the verification of their influence on the overall control stability. It also permits the consideration of the physical values involved in the actuator control. For instance, modules modelling electrical DC motors allow the application of the motor voltages to the controlled system.
- Last, a module simulating the behaviour of a feedback controller has been introduced in the dynamic analysis of ADAMEUS. It elaborates the control order to apply to the actuators from the sensor measurements and a desired trajectory. It can simulate continuous-time controllers, sampled-time controllers and controllers featuring three different stages working at different sampling rates. The simulated controller is configurated in the Preface by reading a fixed-format control data file, that can be released by any software for Computer Aided Design of Control.

3.3. Computation of Inverse Dynamics

The most general control schemes that have been proposed for controlling mechanical systems such as robots all propose a two-stage architecture:

- A nonlinear feedforward aiming at minimising the tracking error between the desired trajectory and the real system behaviour. The most accurate feedforwards ususally rely on the dynamic model of the structure.

- A linear (possibly time-varying) feedback, that corrects the deviations due to modelling errors.

Thanks to its general and systematic generation of the equations of the dynamics, ADAMEUS is able to compute the model dependent parameters of a control law. Such an ability has been demonstrated in SACODY, during which the ADAMEUS capabilities have been extended in order to compute the nonlinear inner loop of the generalized nonlinear decoupling control method, presented Chapter 2, Section 4 in this volume (equation 19).

4. Application of ADAMEUS for Modelling an Industrial Robot

4.1. Brief Description of the KUKA Robot IR 160/60

Like all the major outcomes of the SACODY project, ADAMEUS has been applied to the project industrial demonstrator. It was indeed necessary to demonstrate the relevance of the overall robot control design methodology proposed in the project, and especially that of dynamic modelling for simulating and helping the control design of a non-rigid robot.

In this last phase of proof-of-concept, SACODY focused on the case of the industrial robot IR160 commercialized by the german robot manufacturer KUKA, member of the SACODY consortium.

The robot IR 160/60, shown in figure 2 is a six-axis robot with articulated kinematics, mainly used in spot welding applications. It weighs *1.8 tons* and is designed to carry *60 kg* payloads over ranges reaching *2.5 meters*. Despite very stiff arms and a very heavy structure resulting in a *3.5 %* payload-to-weight ratio, this robot suffered from vibrations, due to elasticities located in its joints when submitted to high motor accelerations. This motivated its choice by KUKA as a demonstrator for the SACODY methodology.

The payload and the dead weight of the articulated components are statically compensated to a large extent by a closed hydraulic counterbalancing system, which assists the second axis. All the axes are driven by electric DC motors. The robot to be modelled carries a *45 kg* spot welding gun, at the end of an arm featuring a 400 mm extension.

4.2. Objectives and Modelling Trade-offs

The objectives that were assigned to the robot model to be developed were:

- To demonstrate as far as possible the different functionalities of the ADAMEUS software,
- to provide the background data for the analysis of the flexible behaviour of the robot and for the design of its antivibration control,
- to allow the verification of the new control by means of numerical simulations.

Figure 2: The Robot KUKA IR160/60

4.2.1. EXPERIMENTAL BACKGROUND

A modal analysis campaign carried out on the robot by the company LMS (see this volume chapter 3) had enabled the identification of five robot vibration modes at frequencies lower than *25 Hz*. Out of these modes, only the first two vibration modes appeared to be significant in terms of vibrational behaviour of the robot. They exhibit frequencies lower than *10 Hz* in all the robot working envelope.

The first flexible mode (see figure 3) is a rotation of the robot structure around a vertical axis due to a flexibility lumped in the first robot joint.

A modal analysis with external excitation and motors on brakes allowed the measurement of a vibration frequency varying from *5.2 Hz to 7.3 Hz* depending upon the robot configurations.

The second flexible mode (see figure 5) is a vibration of the robot arm in a vertical plane due to a flexibility lumped in the second robot joint.

The modal analysis with external excitation and motors on brakes identified for this mode a vibration frequency varying *from 7 Hz to 8.3 Hz* with the robot configurations.

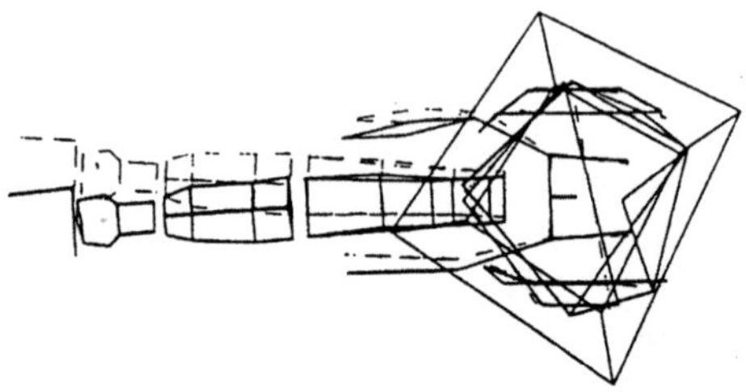

Figure 3
First flexible mode of the robot (top view)

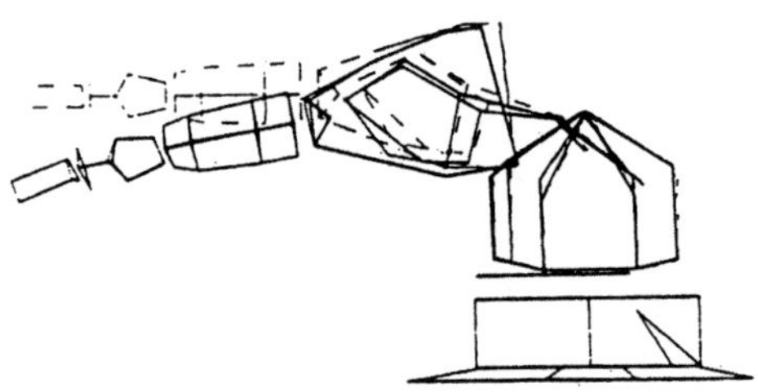

Figure 4
Second flexible mode of the robot (side view)

4.2.2. Description of the Model

57 elements have been used to model the whole robot, two of these elements being flexible shafts located in the gearings in order to reproduce the two first spring-like joint behaviour identified during the measurement campaign.

Model of the Gearings of the First Axis

The rotational motion of axis 1 is composed of a DC motor attached to the fixed base and a series of four speed reduction stages.

The four reduction stages have been modelled with ADAMEUS. Moreover, a flexible element has been introduced in the gearings in order to reproduce the spring behaviour of the first joint identified during the test campaign. 14 different bodies are used to describe the motor and the different gears of the first axis.

Model of the Rotating Column

The rotating column is composed of:
- the rotating column itself,
- the motor driving the second axis,
- the elements of the gearings of axis 2 (three reduction stages),
- the hydraulic counterbalancing system for compensation of the payload and the dead weight of the articulated components.

The shaft located in the gearings has been modelled by a flexible element in order to reproduce the spring behaviour of the second joint identified during the test campaign. The hydraulic system is modelled with a linear actuator. 12 different bodies are used to describe this rotating column.

Model of the Link Arm

The link arm is made of:
- the link arm column itself,
- the motor driving the third axis,
- the elements of the gearings of axis 3 (three reduction stages).

10 different bodies are used to describe this rotating column.

Arm, Wrist and Tool

The arm is the most sophisticated part of the IR 160 KUKA robot. Since no flexibility in this arm was exhibited by the experimental tests, the set of gears of the wrist axis has been simplified in the model in order to reduce the amount of degrees of freedom and consequently computation time. 21 bodies were used to model the arm including motors and gears and the tool.

4.2.3. Remarks

Flexible elements

The flexibilities have been modelled by two elastic bodies, located in joints of axis 1 and 2, and characterised by means of a Nastran Finite Elements analysis. The files resulting of this Finite Element analysis, containing the mesh, eigenpulsations, generalised masses & stiffnesses and mode shapes of the flexible bodies, have been read by ADAMEUS when processing the robot model bulk data.

The complete models of the main axis gearings appeared in fact absolutely necessary to reproduce accurately the evolution of the vibrations as a function of the robot configuration. 17 gears have thus been included in the complete model.

Closed mechanical loops

Four closed mechanical loops are present in the model, due to the counterbalancing system and to the complete description of the gearings of the three main axis.

Reduction of degrees of freedom

After consideration of the motion constraints, ADAMEUS retains 45 working degrees of freedom. Out of these 45 working degrees of freedom, only 8 effective degrees of freedom are retained, namely the six motor rotations and the two joint deflections.

Figure 5 shows the graphic representation of the robot in its definition configuration.

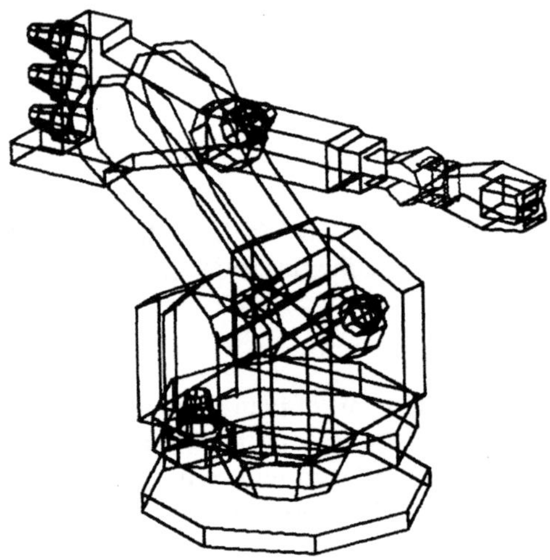

Figure 5
Definition Configuration

4.3. Origin of the Model Data

Geometric data such as:

- coordinates of GRID points (joints),
- local coordinates systems,

have been defined after a general drawing made available by KUKA. Values were given by KUKA for the masses, positions of mass centers and the main inertia of the three main components of the robot. Reducing ratios and shaft rotation inertia have been specified by KUKA as well as the characteristics of motors. Values for the flexibilities (location, stiffness, damping) have been defined after combined study of the driven inertia and of the results of the identification campaign.

4.4. Validation Results

4.4.1. Comparison with the Results of Modal Analysis

In order to validate the model, its modal analysis has been carried out by linearizing the equations of the robot around the three test configurations used by LMS during their experimental campaign, which provide a good assessment of the maximum variation of the robot inertia.

The frequencies found after analysing the model modal behaviour are compared in table 1 with the values measured for the two first modes during the test campaign. This table shows that the discrepancies between the model and the experimental frequencies remains below 3 % in the three test configurations.

	Experimental test	*ADAMEUS model*	*Modelling Error*
Test Configuration 1	*5.17 Hz*	*5.17 Hz*	*0 %*
Test Configuration 2	*7.19 Hz*	*7.19 Hz*	*0 %*
Test Configuration 3	*7.25 Hz*	*7.42 Hz*	*+ 2.3 %*

Frequency of the first mode (Torsion in the first joint)

	Experimental test	*ADAMEUS model*	*Modelling Error*
Test Configuration 1	*7.05 Hz*	*7.05 Hz*	*0 %*
Test Configuration 2	*8.31 Hz*	*8.54 Hz*	*+ 2.7 %*
Test Configuration 3	*7.98 Hz*	*8.01 Hz*	*+ 0.4 %*

Frequency of the second mode (Torsion in the second joint)

Table 1
Comparison Between the Model and the Experimental Modal Analysis

4.4.2. Open-Loop Simulation and Comparison with Experiments

In order to validate furthermore the model derived, the open loop behaviour of the robot has been compared with that of its ADAMEUS model submitted to the same torque command.

The experience has been performed on the first axis of the robot in an extended configuration. Different motion records have been compared with the results obtained in simulation of the same experience.

Figure 6 compares the horizontal tool accelerations obtained in experience and simulation in response to the same open-loop torque profile. This simulation shows that the model enables an accurate reproduction of the robot dynamics, including its vibrational behaviour.

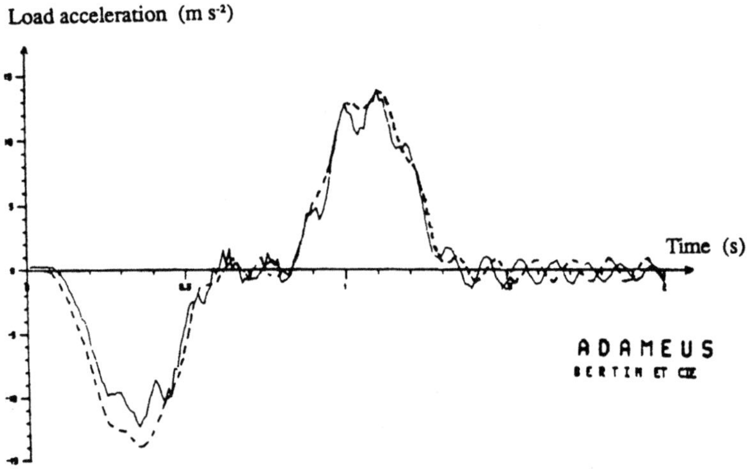

Figure 6 Comparison between simulated (dashed line) and experimental (solid line) robot end-point acceleration under the same open-loop torque command.

4.5. ADAMEUS for Control Design and Closed-Loop Simulation

This section illustrates the capabilities of ADAMEUS as a tool for designing the control of multibody servomechanisms. The procedure described is also shown on the flow chart figure 7.

Once a complete model has been derived, the first task consists in performing the analysis of the system to be controlled. To this aim, linear models of the robot are derived by means of ADAMEUS. The resulting file contains for each prescribed linearization points, which in this case consist only of the first LMS configuration test, the mass, damping and stiffness matrices of the equivalent linear systems, as well as the corresponding matrices A and B of the corresponding augmented state-space equations.

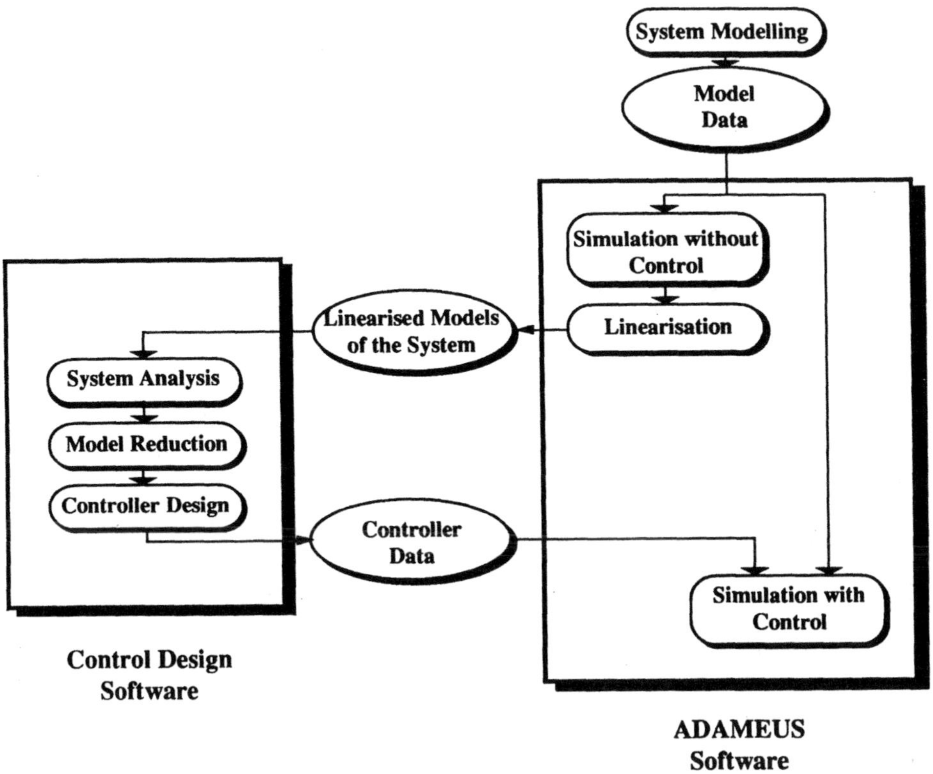

Figure 7
Procedure for designing the control of flexible servomechanisms modelled with ADAMEUS

This file can then be read by any Computer Aided Control Design software, like for instance MATLAB, MATRIX-X or as it has been tested in this case, BASILE. Such software can then be used to engineer the control feedback on the basis of the simplified model, and to generate a control data file to be used by ADAMEUS to simulate the behaviour of the controlled system.

The control architectures (Figure 8) that can currently be simulated by ADAMEUS may feature:

- An analog (continuous-time) servo level,
- A Digital servo level working at a sampling period specified in the data,
- A Trajectory generation stage, reading in a trajectory file the trajectories to be tracked by the system, and releasing it to the digital servo level with a specified rate, usually superior to that of the digital servo.

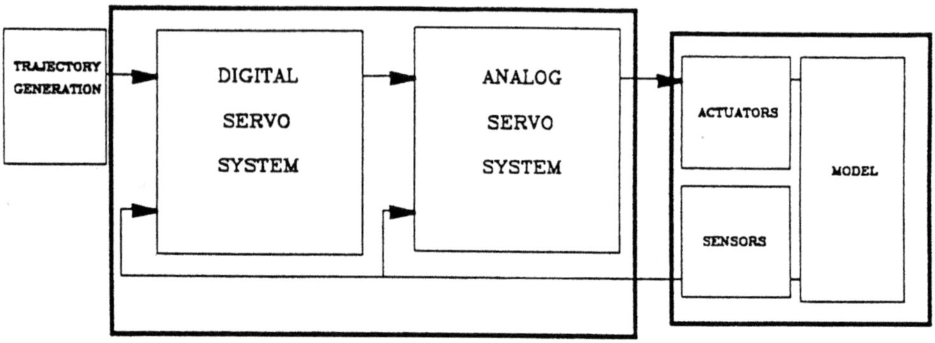

Figure 8: Control structure simulated by ADAMEUS

Figure 9 shows the results of a closed-loop simulation performed on the first axis of the robot KUKA IR160. For this simulation, the digital servo is computed each 6 ms and the trajectory is read each 48 ms.

5. Conclusion

This paper has presented the development carried out in the framework of SACODY in the field of dynamic modelling of non-rigid multibody mechanisms.

The general purpose computer code computer code ADAMEUS, the first commercial version of which was made available in 1988, has been enhanced by the introduction of new functionalities required for designing the control of lightweight robots:

- possibility of simulating controlled systems,
- ability to derive linearised models,
- ability to derive control parameters depending on the model, like dynamic feedforward.

These new functionalities should be integrated in the next software version, foreseen in 1993.

Other developments concerning the simulation of hydraulic actuators are currently under way.

With the completion of these developments, a generic tool for the dynamic simulation of non rigid servomechanisms will be available. It is already suited for the design of complex control suchlike the SACODY antivibration control. In a near future when lightened structures and advanced control are simultaneously engineered to take full benefit of the SACODY results, ADAMEUS should be a powerful tool for optimising the design of lightweight servomechanisms.

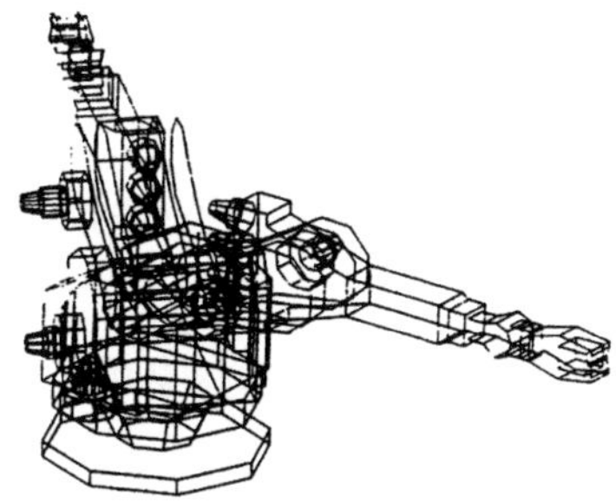

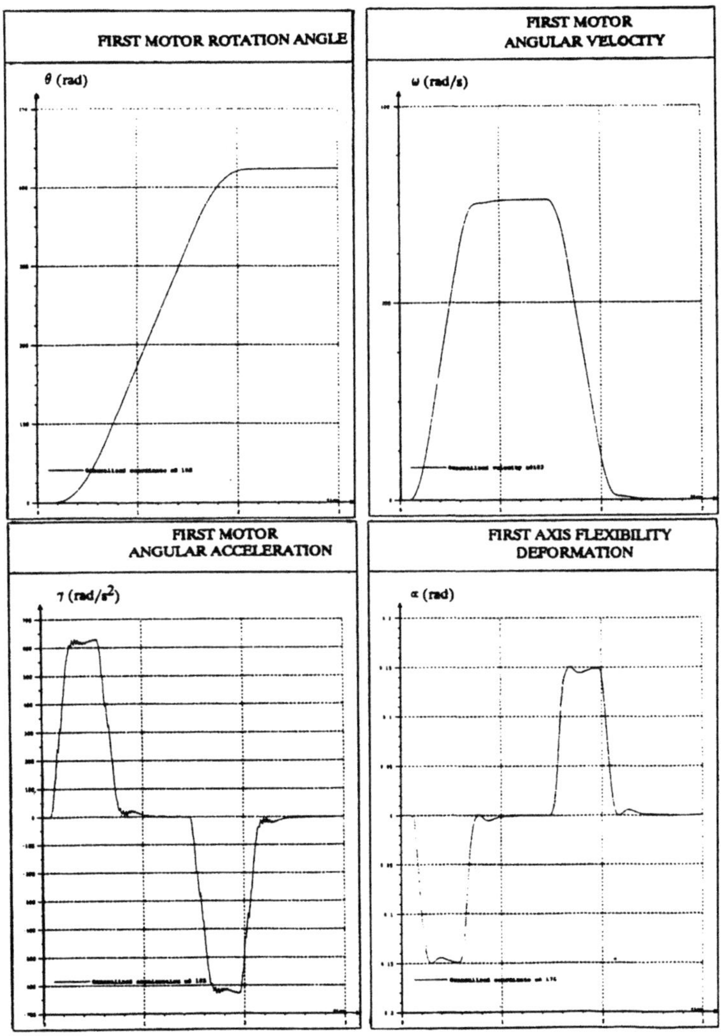

Figure 9: Closed-loop simulation by ADAMEUS of the robot IR 160

Bibliography

[1] Kane T.R., "Dynamics of Non Holonomic Systems", Journal of Applied Mechanics. 28. pp 574-578, 1961.

[2] Huston R.L., Passerello C.E. "On Lagrange's Form of d'Alembert Principle" Matrix and Tensor Quart, 23, 1973.

[3] Huston R.L., Passerello C.E., "Multibody Structural Dynamics including Translation between the Bodies", Journal of Comput. Struct. 12, pp 713.720, 1980.

[4] Chace M.A., Bayazitoglu Y.O., "Development and Application of a General d'Alembert Force for Multifreedom Mechanical Systems" Journ. Engin. Industr., 93, pp 317-327, 1971.

[5] Willems P.Y., "Attitude Stability of Deformable Satellites", Colloque International sur la Stabilisation des Satellites CNES, 1968.

[6] Fraeijs de Veubere B., "Non Linear Dynamics of Flexible Bodies", Symposium on Dynamics and Control of Non Rigid Spacecraft", ESA Frascati, 1976.

Chapter 5

CAD/CAE based Methodology for Optimising Robot Mechanical Design

Barry Duignan, Pat Connolly
University College Dublin

1. Introduction

This report outlines the work done at University College Dublin under the General heading "Mechanical Design and Optimisation". The main part of the work outlined here is the development of a design and optimisation strategy for industrial robots, integrating the various Mechanical Computer Aided Engineering (MCAE) methods. The strategy, once formulated, was demonstrated using design data for the KUKA IR 160/60 robot. Also in relation to this work, robot simulation software was written to investigate trajectory control in the case of robots requiring redundant degrees of freedom.

2. Background

2.1. Background within SACODY

One of the main criteria of robot design is positional accuracy, and given adequate sensors and control, the main factor which will degrade accuracy is vibration. Lack of stiffness and high speeds both contribute to vibration, and the traditional way to avoid vibration is to design the robot to be structurally stiff enough to ensure that vibrational modes are not excited and that static deflection is not excessive. The aim of the SACODY project is to actively control robot vibration, so that the robot can be driven harder without vibration becoming a problem.

This raises a number of questions relating to the design of the robot. Most importantly, the design of robot elements now becomes the task of tailoring the structure to obtain optimum performance with vibration characteristics suited to the controller. This is a difficult task, and the extensive prototyping and testing required can be substantially reduced by the use of MCAE techniques both to design the components and to simulate their operation in situ. In order to achieve this, the two separate technologies of Finite Element Analysis (for vibration characteristics) and Dynamic Analysis (for mechanism simulation) must be used together in an effective way. On a broader level, the integration of these technologies with workspace analysis methods allows the integrated CAE strategy to be used to the full from the concepts stage to the final design, allowing a greater range of concepts to be examined and reducing prototyping.

2.2. General Background

These tools have not heretofore been widely used in the robot design process, but recent developments in the computer industry have produced high-powered workstations with powerful graphical abilities. This means that engineering analysis packages no longer need mainframe power to be practical for complex problems. Also, the increased graphics capabilities of workstations mean that these packages can be much more user-friendly - the prospective user no longer needs to become an "expert" in the particular package to use it effectively.

The adoption of UNIX and X-windows has also led to packages being more compatible and flexible, since packages can be run remotely and provide a full graphical interface to the user locally if both machines run X.

These factors have pushed packages together so that, for example, a part designed using a solid modeller could be easily analysed using a Finite-Element analysis (FEA) package by having an element-mesh generator incorporated in the modeller. Another example of this kind of cooperation is between FEA packages and Dynamic modellers, which can use FEA ouput to model flexible bodies.

This integration is the key to fuller utilisation of these packages. The I-DEAS package from SDRC is a very good example of this, with a solid modeller, a drafting package, a finite element package and a dynamic modeller all working from the same database.

2.3. Types of Analysis Used

In this section, the four main types of analysis used in the strategy are discussed. Briefly, these are Kinematic analysis for examining the layout of the robot (number and types of joints and links, etc.), Dynamic analysis for simulating the forces and accelerations of the robot in motion, Finite Element analysis for stress and vibration analysis, and Modal analysis of actual structures, to confirm simulated results.

2.3.1. Kinematic analysis

Kinematic analysis examines structures from a purely geometric point of view, thus avoiding questions relating to the masses, inertias and forces. In the initial stages of robot design, basic questions of how many joints to have, what type (i.e. rotating or sliding) and what dimensions the links should have are best answered using kinematic analysis. Of course, some consideration of the dynamics involved must be made, and realistic assumptions of maximum speeds and accelerations must be borne in mind at this stage.

Kinematic analysis takes two forms: Workspace analysis and direct simulation. Workspace analysis involves the parameterisation of the geometrical equations describing the robot's structure. These parameters can then be used to predict the size and quality of the robot's working envelope, and the locations of any degenerate configurations (that is, positions *within* the working envelope from which motion in certain directions is impossible).

The second form of kinematic analysis, implemented in this work, is direct simulation. In this method, the equations relating each joint with the subsequent one are used as the basis. From these, equations giving the joint velocities required to keep the tool point along a prescribed trajectory can be worked out. By using these equations in a generalised matrix form, almost any robot geometry can be modelled.

2.3.2. Dynamic Analysis

Dynamic analysis is the simulation over time of mechanisms. The dynamic analysis package consists of three parts: a pre-processor, which produces the dynamics equations from user input describing the robot in terms of bodies links, constraints, forces, etc.; a solver which solves the equations over time; and a post-processor to permit viewing of the results of the simulation in graphical or other format.

In robot design, dynamic analysis allows a model of the robot to be built up, body by body. This model can then be analysed in two ways: Either by applying forces and torques to the bodies representing the motors and observing the performance of the robot, or by specifying a path for the tool to follow and computing the forces and torques required. Dynamic modellers can now also utilise the results of finite element modal analysis and so model the behaviour of flexible bodies.

Dynamic modelling therefore has three roles to play: First, it is complementary to kinematic analysis, insofar as it can allow a more detailed consideration of the impact on dynamics of a particular structure. Second, it plays a very useful role in the evaluation and sizing of transmission and actuation elements, by computing the forces borne by them in simulated normal operation. Finally, it allows the simulation of the in-service behaviour of flexible bodies, and their collective effect on robot accuracy and vibration generally important for the consideration of wear and vibration tolerance of components.

2.3.3. Finite Element Analysis

Finite element analysis has two roles to play in this scheme. First, it can be used to obtain the modes of vibration of flexible bodies, be they arms or shafts or other drive elements, and produce the results in a form usable by the dynamic modeller. Second, it can be used for stress analysis and weight reduction. Since many packages now feature weight optimisation algorithms, finite element analysis can be used to minimise the weight of bodies subject to criteria of vibration frequencies, static deflection and stress.

2.3.4. Modal Analysis

Modal analysis can be used to extract vibration and deflection information from the actual robot, as demonstrated by the work done using LMS's CADA-X software. This information can be used to reinforce and confirm the results of simulations. With the recent advances being made in software user interfaces, this information can also be examined in parallel with simulation information, and thus compared directly.

2.4. Application to Robot Design

In summary, the four technologies mentioned above interface in the following ways:

The Kinematic Analysis is used to look at the purely geometrical aspects of the design. In the design phase, this would be:

- Choosing the basic configuration (number of axes, arm lenghts, wrist type, etc.)
- Optimising the design to maximise the working envelope and manoeuvrability as well as to minimise or eliminate the occurence of degeneracies and singular configurations.
- Examining the possibilities of using redundant degrees of freedom to enhance performance.

The Dynamic Analysis is used at the design stage to:

- Specify motors and transmission types, given an initial structural design.
- Simulate and compare design alternatives under similar conditions of usage.
- Evaluate and modify vibration characteristics of drive and structural elements (with the aid of modal analysis).
- Maximising performance by reducing weight and increasing motor power as much as possible while avoiding excessive vibration and deflection problems.
- Producing plots of forces on various elements over time.

The Finite Element Analysis provides modal analysis and stress analysis for:

- Obtaining vibration behaviour for structural elements in a form eadily usable by dynamic modeller.
- Optimisation of robot arm from points of view of stiffness (both vibration and static deflection) versus weight.
- Stress checking using force data from Dynamic analysis.

The Modal Analysis. Sensor input from the actual robot is analysed to extract modal information. Modal Analysis is useful for existing or prototype robots for:

- Identification of modes of vibration
- Verification of any design changes implemented.

The diagram Figure 1 shows how the integrated CAE approach works from the initial design stage of a robot, with the initial definition of the robot market niche to the fine-tuning of the robot's vibration characteristics to suit the particular set-up i.e. whether the controller will be capable of controlling certain flexibilities or whether the robot is to be, for all practical purposes, rigid.

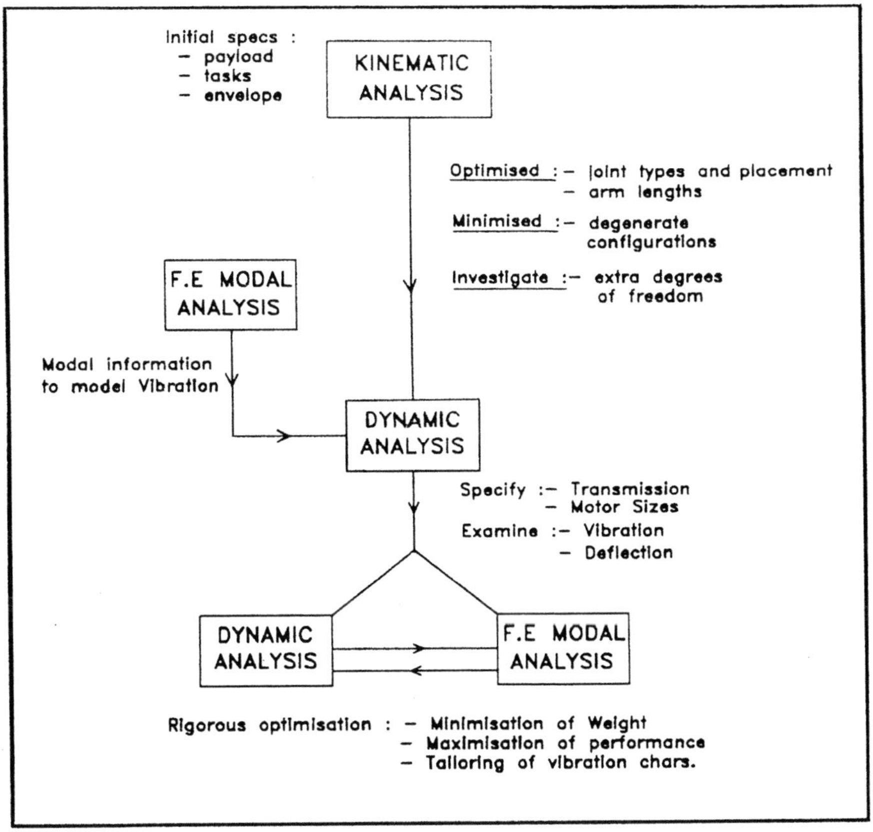

Figure 1
The full implementation
of the integrated design strategy

The diagram Figure 2 shows how increasing the torque of a motor (or motors) is examined. Again, this loop is iterated until an optimum balance between motor torque, motor weights, increased vibration and loads on components affected by the increase in torque.

The chart Figure 3 shows how the Modal and Dynamic analysis can be used to analyse the effects on performance of reducing arm weight. Also the effects on the arm in terms of vibration characteristics, deflection under load and stress can be examined. The above loop is iterated until an optimum design is obtained.

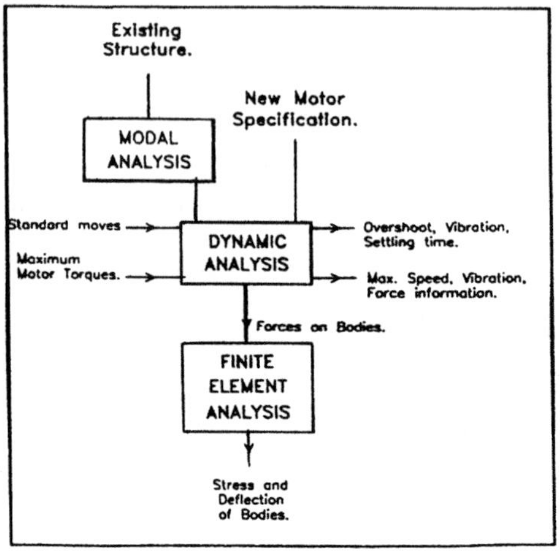

Figure 2
Flow of information for analysis of increased motor size

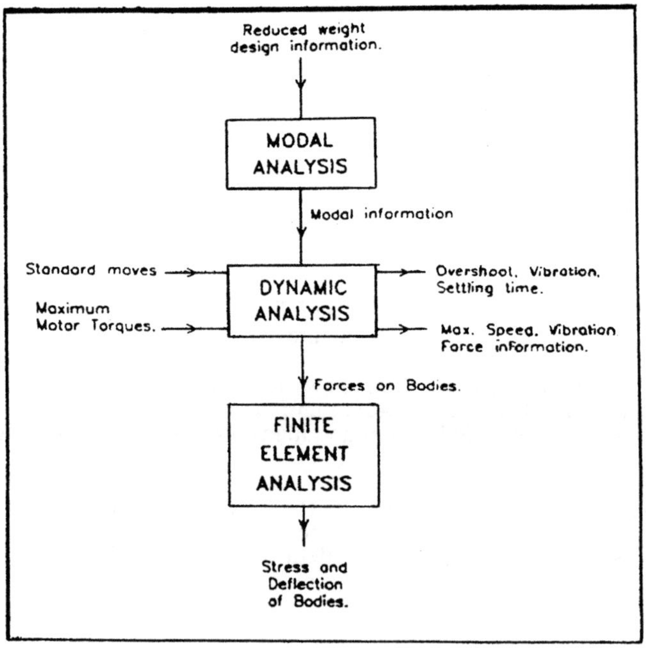

Figure 3
Flow of information for a component weight reduction analysis

3. Software Used

3.1. Overall software

This section outlines the software used in the formulation of the design scheme.

3.1.1. Kinematic Analysis

For kinematic analysis, a generic graphical simulation package was used which was developed at UCD for the task "Trajectory Control with Degeneracy". This program, described in the next section, was initially written to examine mathematical methods of coping with the problems of degeneracy. It requests the user for motion commands and simulates the resultant trajectory by displaying three orthographic views of a wireframe model of the robot and listing the instantaneous tool position to both the screen and datafiles for subsequent plotting. The package was also used to prepare joint motion data for a simple model of the KUKA IR 160 created on the DADS dynamic modeller, described below.

3.1.2. Dynamic Modelling

Two dynamic modelling pakages were used in this work. DADS, from CADSI, Inc. was used in conjunction with the kinematics package above, and ADAMEUS, developed by BERTIN, was used for the major part of the work.

3.1.3. Finite Element Analysis

For finite element analysis, ANSYS, from Swanson Analysis was used for the transfer of finite element data to ADAMEUS and for initial examination of the structure of the KUKA robot arm. ABAQUS, from HKS was also examined for these tasks. The main part of the analysis work, including the optimisation of the arm, was done using I-DEAS from SDRC. I-DEAS is a suite of analysis and design programs, based around a solid modeller. In this way, drawings, FE meshes and mechanisms can be developed from an initial solid model design.

The optimisation element of I-DEAS works by minimising the mass of an object subject to certain constraints. So for example, an object (such as a robot arm) modelled using shell elements would have the thickness of these elements reduced until the user-defined constraints (stress, displacement, modes of vibration, etc.) were reached.

3.2. Hardware Implemented

Software product	platform	specifications
I-DEAS level V	DECstation 5000	24 MIPS, 24 MB RAM, 3.7 Megaflops
ADAMEUS	Sun Sparcstation 1 +	16 MIPS, 8 MB RAM, 1.7 Megaflops
ANSYS	VAX 11/780	1 MIPS, 14 MB RAM,
Kinematic simulation	VaxStation 3100	2.5 MIPS, 16MB RAM

3.3. Kinematic Simulator

The kinematic simulation software was developed for the SACODY task entitled "Trajectory Control with Degeneracy". It was written in Fortran 77 on a Vaxstation 3100, with a graphical interface using GKS. It runs under both the VWS and DECwindows interfaces and can provide hard-copy output in HPGL and Postcript.

The program simulates straight-line robot tool trajectories using a parabolic acceleration-deceleration profile as suggested by the Katholieke Universiteit Leuven. Any robot having up to eight axes (either rotating or sliding) can be simulated. The equations used are matrix-based and modular in nature, and so the program can be extended to any number of axes.

3.3.1. Degeneracy

In order to fully describe the position and orientation of an object, six coordinates are required. Three for position and three for orientation. This means that a robot will need six axes of motion to be able to fully position and orient a tool. In some configurations, however, the robot loses tha ability to move in certain directions. For example, if two axes lie along the same line, moving either will have the same effect.

This must mean that there are some directions along which motion is impossible. Also, even if the desired motion is physically possible, the fact that two axes provide the same motion means that the kinematics equations have multiple solutions (i.e. the desired tool motion may be achieved in more than one way). This may cause them to break down. Such configurations are known as *degenerate*.

Degeneracy therefore gives rise to two types of problem: the first is physical, motion in some directions being impossible from degenerate configurations. The second is mathematical, where degenerate configurations cause numerical problems with kinematics equations.

The solution to the first type of problem is to ensure that the robot possesses no configuration from which motion is impossible. This is most easily done by adding a seventh redundant axis of motion. This means that almost every tool motion can be achieved in more than one way and measures outlined below are needed for the kinematics equations. Also, if deviation from the specified path is permissible in certain directions, the kinematics equations may be written so as to produce a solution which optimises the deviation necessary.

The solution to the second type of problem is to find an optimum solution from the many available, based on a criterion such as minimum kinetic energy or minimum joint motion.

3.3.2. Trajectory Control

In order to control the trajectory of a robot operating in Continuous-Path (CP) mode (i.e. following a particular path through space) it is necessary to compute the incremental motion of each joint required to produce the desired increment in tool position at each controller cycle. If the position and orientation of the tool is described by the vector x and the positions of the joints are described by the vector θ then a trigonometrical function relating tool position to joint position can be written:

$$x = f(\theta) \tag{1}$$

To obtain the joint positions required for a particular tool position, equation (1) must be inverted following:

$$\theta = f^{1}(x) \tag{2}$$

To obtain the incremental joint motions required for incremental tool motions, the basis of CP motion, equation (2) must be differentiated. Equation (2), however can be difficult or impossible to obtain from equation (1) for robots having six or more axes. By arranging axes of motion to be parallel or perpendicular, this can be done, but since this package was to be as general as possible, a general algorithm was used. Therefore, instead of relying on obtaining the derivative of (2), the derivative of (1) was first obtained:

$$\Delta x = J\Delta\theta \tag{3}$$

And then inverted:

$$\Delta\theta = J^{1}\Delta x \tag{4}$$

This equation is now always explicitly obtainable in matrix form. J is the function relating joint velocities to tool velocities and in (3) above is represented by a matrix of partial derivatives of (1). The solution of equation (4), therefore can be seen as the inversion of the matrix J, so the degeneracies correspond with the matrix J being singular (i.e. no inverse exists).

3.3.3. Behaviour of the Matrix J

There are three ways in which equation (4) can behave. There can be no solution, a single solution or many solutions. A single solution is the normal situation for six-axis manipulators - the desired increment in tool position can be obtained by one set of incremental joint motions. Many solutions mean that the desired incremental motion can be carried out by more than on set of joint motions. No solution means that the desired motion is physically impossible.

In the case of there being multiple solutions, it is desirable to find the optimum solution based on some criterion. For example the joint motions which would minimise kinetic energy (joint acceleration) would reduce power consumption, wear and vibration. In the case of no solution, it might be desirable to produce a trajectory for which the error is minimised, or for which the error is allowed to be greater in certain directions than others.

3.3.4. The Pseudoinverse

The pseudoinverse of a matrix is defined by minimising a velocity-squared criterion using Lagrange multipliers with equation (3) above as a criterion. The result, as applied to the kinematics problem, is an inverse function which always exists, even when the classical inverse is singular. When the classical inverse produces no solution, the pseudoinverse produces an approximate solution with a least-squares error, and in the case of multiple solutions, the pseudoinverse produces the solution requiring least kinetic energy. Several methods of implementing Pseudoinverse-based solutions were investigated using this software.

The diagram next page shows the flow of information between the Kinematic Analysis main program and the output files and display.

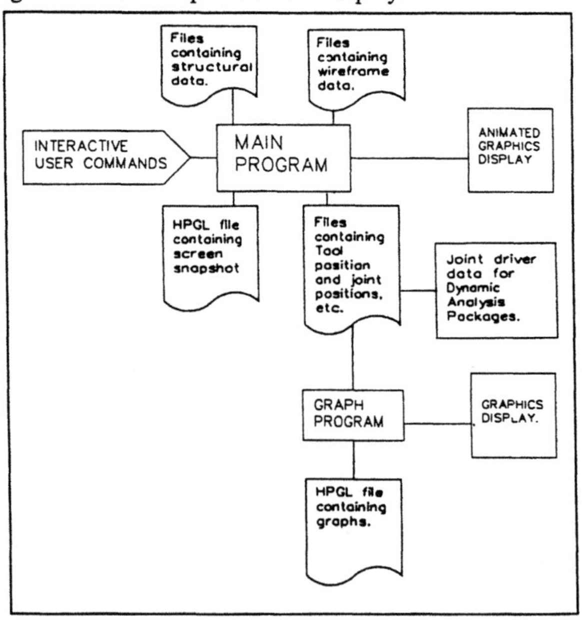

Figure 4
Organisation of the Kinematic Analysis Software

The graphic display shown figure 5 consists of three orthographic views of the wireframe model, together with text describing tool position and joint positions. The tool position, and joint positions, velocities and accelerations are written to data files at each timestep. These files can be reviewed using the graph program, which plots to the screen and, if required, to a HPGL plot file. In addition to this, a "snapshot" showing the wireframe model at some specified time can also be written to a HPGL file from the main program. This will plot an exact copy of the main program's screen output at that instant.

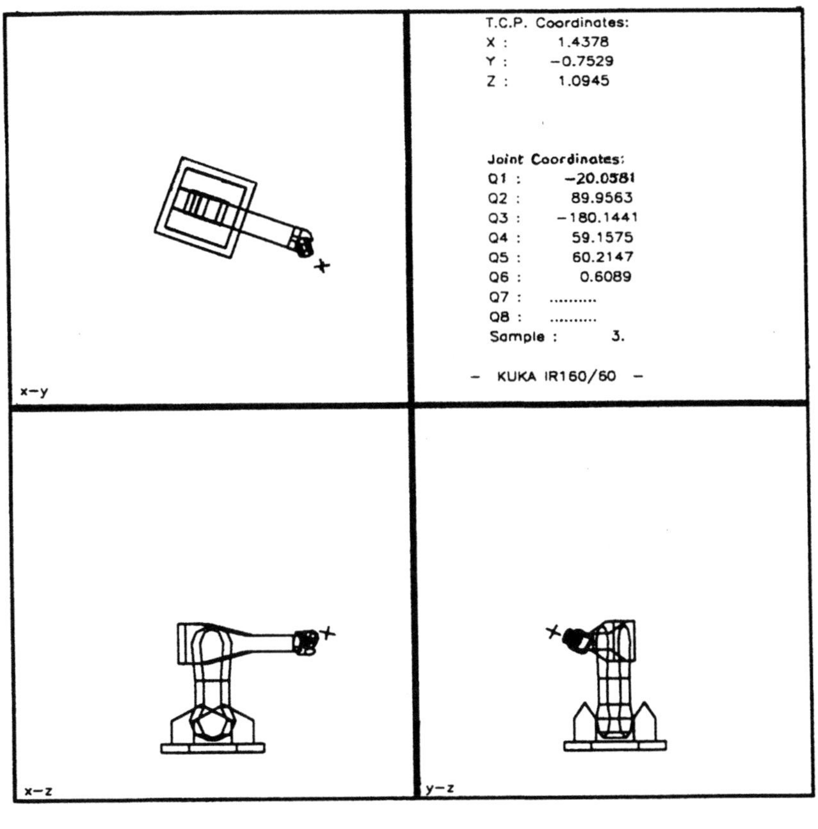

Figure 5
Graphical output of the main program
as seen interactively by the user.

4. Arm Weight Reduction Analysis

The primary subject of this section is the integration of different CAE tools. For the main part of the demonstration of the integrated design scheme, therefore, the reduction of section thickness of the inner arm of the KUKA IR160/60 has been chosen, since it incorporate finite element analysis and optimisation, dynamic modelling and solids modelling.

For the first part of this work, a solid model of the arm was created on the I-DEAS solid modeller. This model was used to obtain surfaces on which meshes of shell elements were applied.

The model was constructed entirely of shell elements, and restrained at the points of attachment to the carousel. The second arm, wrist and maximum payload were represented by a lumped mass-inertia element, attached to the arm by two rigid elements. The other parts of the arm associated with the drives, etc. were ignored since their mass and inertia is small compared with that of the rest of the robot, and their stiffness (i.e. belt drives) is lower than that of the arm.

This model of the arm was analysed using the I-DEAS FE solver. Two runs were made; the first was a modal analysis, the second was a static analysis. The modal analysis yielded a natural frequency of *18 Hz* and the static analysis, simulating maximum accelerations of axes 1 and 2, with the arm in an oustretched position, yielded a maximum deflection of *1.5 mm*.

The optimisation of the arm consisted of reducing the thickness of the two sections shown in figure 6, subject to a constraint of keeping the natural frequency above *16 Hz*.

The reduced weight information was then fed into the ADAMEUS model, and the improvement in performance examined by accelerating the robot at its maximum torque over 1 second (the acceleration time as per specifications).

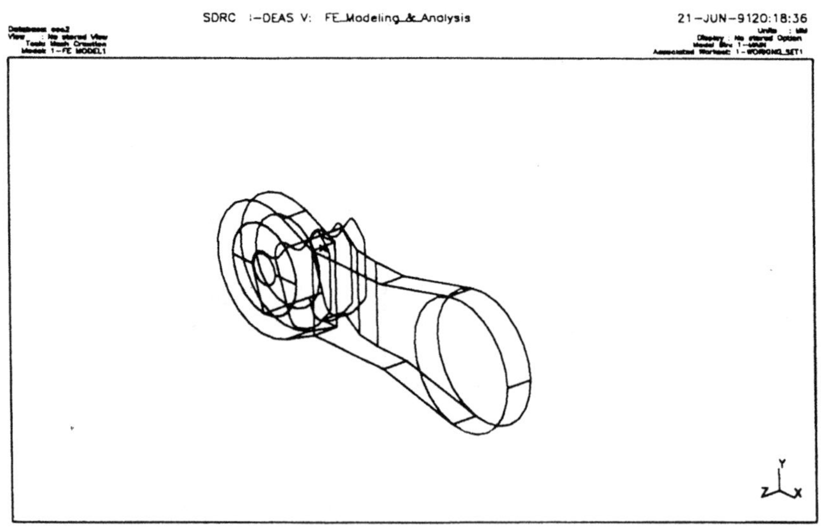

Figure 6: Mesh areas derived from the solid model.

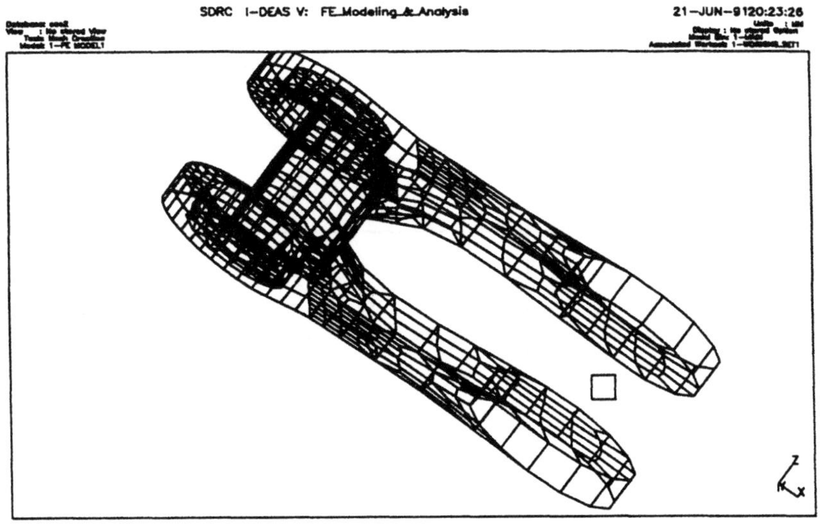

Figure 7
Shell element model of the arm.
The diamond-shaped item between the sides shows the location of the mass representing the rest of the robot.

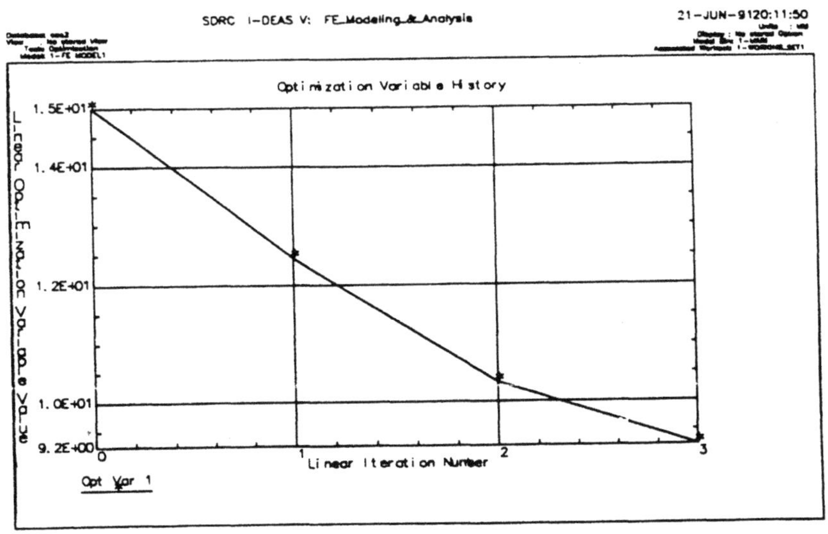

Figure 8
Thickness of the element at each of the three optimisation iterations, starting at 15 mm and reaching 9.425 mm approximately before reaching the target frequency of 16 Hz.

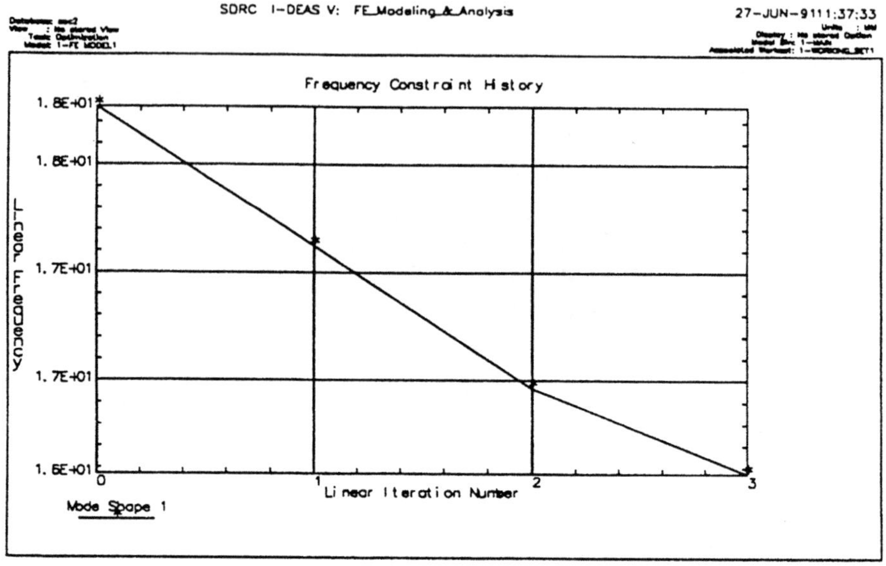

Figure 9
Natural frequency of the arm at each stage of the optimisation.

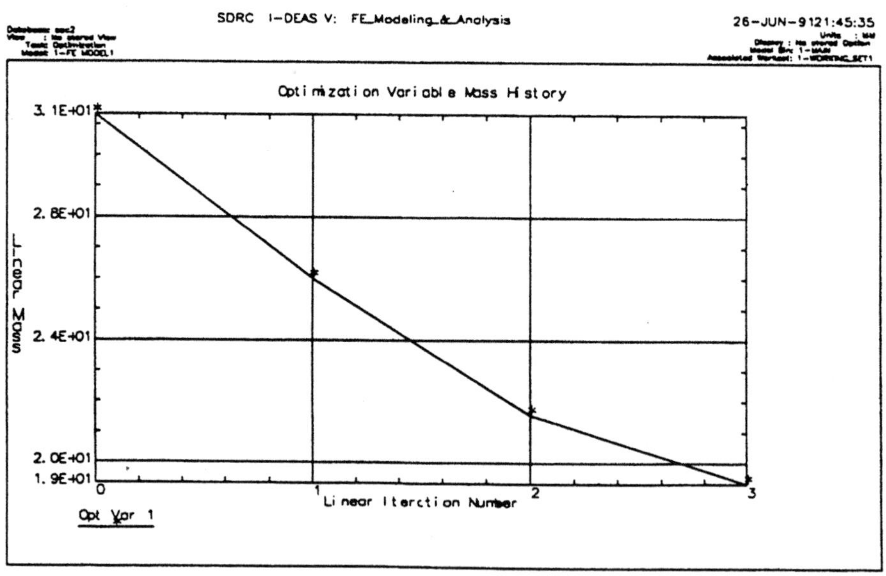

Figure 10
Mass of the part of the arm subjected to optimisation at each stage.

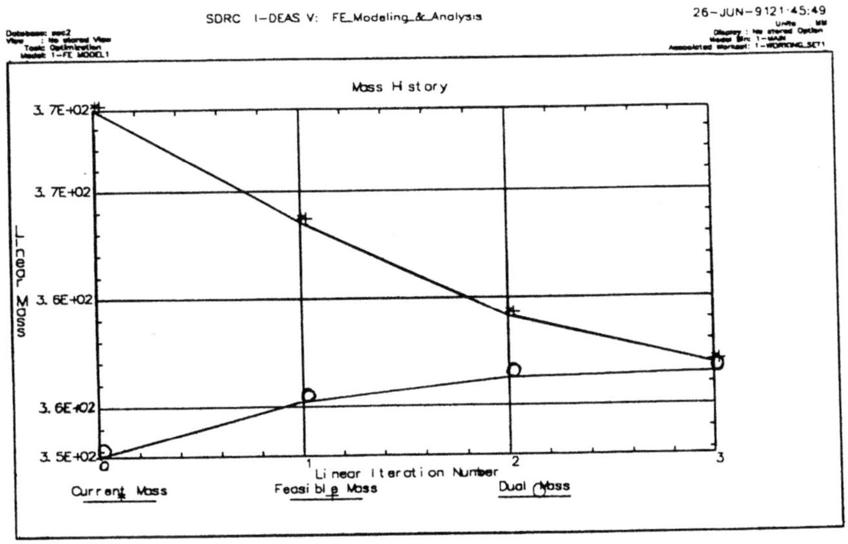

Figure 11
Total mass of the arm structure at each stage of the optimisation

5. Conclusions

5.1. Impact on Machine Design

The tools specified in previous sections allow the designer the greatest possible flexibility in examining design alternatives from the earliest stages. The kinematic analysis package allows the designer to thoroughly evaluate a number of configuration options at a very early stage. The tight coupling of dynamic and finite element analysis provides a means of rigorous analysis of structural and transmission components from the points of view of stress, vibration and performance. The inclusion of test data improves the quality of simulation as well as giving the advantages detailed in the next section.

As well as aiding design, these tools also allow analysis of the effects of upgrading or customising existing machines. The extra demands placed on the robot by upgrading motors or increasing payload. This is particularly useful for robot and machine tool builders who are often involved in special one-off designs.

5.2. Utilisation of Test Data

With the continuing integration of packages and data interchange formats, the testing phase can be much more tightly integrated with the analysis. This allows on the one hand, the analysis and simulation data to benefit from the accuracy of actual test data, and on the other hand, allows the full range of software tools to more fully evaluate the test data.

For example, modal analysis data can be converted to the same format as finite element analysis data, thus allowing their direct comparison using the full processing capability of both software tools. Also, the actual test data can be incorporated into simulation models. This is possible not just for modal data, but also for test data on, for example motor characteristics, and friction.

5.3. Integration with CADCAM

The link between the analysis modules and CADCAM packages is another important part of the integration strategy, although these tools were not included in this project.

The analysis of the KUKA IR 160/60 carried out in this work clearly demonstrates the potential of using an integrated Design/Analysis strategy. By using the solid modelling/drafting element of the I-DEAS suite to prepare the arm mesh, a direct link between design and analysis was created. Also, any design changes produced by the optimisation package can be directly returned to the modeller and updated drawings/CNC offline programs/bills of materials can be easily produced.

5.4. Impact on Design Iteration Cycle Time

A key factor in the success of a scheme such as this one is the power of the hardware available. It is now a fact that the sort of computing power necessary for engineering analysis packages no longer requires expensive mainframe equipment and specialist staff and environments.

In this work, the finite element and solid modelling work was carried out on a Decstation 5000: The finite element analysis of the arm, consisting of over 900 shell elements, took less than 30 minutes. The optimisation runs, consisting of 3 recalculations, took approximately 1 hour. The dynamic analysis of the robot, featuring almost 60 bodies modelled over 100 time intervals was carried out on a Sun Sparcstation 1 + and took 10 minutes. Both these workstations are effectively entrylevel engineering workstations and, in fact, both have been superseded by a factor of at least two in price/performance since the commencement of this task.

5.5. Software Considerations

Different software packages can be integrated at two distinct levels: A single environment can be developed to unite them or data can be transferred via standard neutral interfaces, eg STEP.

With computing times for analysis dropping dramatically, the most time-consuming aspect of the type of work outlined in the methodology is the data entry to the various packages. The solution offered by the I-DEAS suite, where data for all parts of the package can originate as a solid model offers the best level of integration, although the package is large and expensive.

This level of integration is not necessary, since the full geometric data is not needed by all elements of the analysis scheme, and provided a good link exists between the Finite Element modeller and the CAD system, other data interchange can occur on quite a low level - eg ASCII representations as in the links between ADAMEUS or DADS and the FEA packages used in this work.

5.6. Comment on Analysis of IR 160/60

The simulation of the robot using the reduced-weight arm showed a slight improvement in speed when maximum torque was applied to the base motor. The effect on deflection of this reduction was slight, and unnoticeable from the point of view of vibration. This could be improved by applying the weight-reduction algorithm to more of the robot's structure, although the weight of the structural elements in relation to the total mass of the robot should be borne in mind.

Also, this analysis is being carried out on an existing structure. The structure is complex and a retro-analysis such as this one, carried out in a limited time, cannot take into account the details of the design. The same process applied to a prototype robot would clearly speed the design greatly and lead to a better final product.

6. Outlook

It is clear from this work that the design of robots can benefit greatly from the type of design methodology outlined in this work. Also, the task of machine element and mechanism design generally can benefit.

Integration of data is the key to the success of the scheme. This has been achieved at a low level in this work by writing computer programs for exchange between Dynamic Analysis and Finite Element analysis.

The ideal level of integration is a single CAD-based environment such as the I-DEAS environment, where the same data is easily transferrable between modules of the package using the same database and "look-and-feel". Clearly, this is made much easier as worldwide developments in standards such as STEP and the OSF proceed, but until then, a lower level of integration must be used, and is possible.

This would allow the individual users to combine specialised packages (such as a kinematics or workspace analysis package) with more common "off the shelf" CADCAM and analysis packages using the most appropriate interfaces. In this way, the user can produce a design system which most satisfies their needs. As standards improve, the task of integration of packages will become easier, just as the adoption of X windows has made user interfaces easier to develop and more portable.

Chapter 6

Sensor Systems for Robot Evaluation

Hendrik Van Brussel, Francky Demeester,
Johan Van den Bossche, Paul Vanherck

K.U. Leuven

1. Introduction

1.1. Problem statement

Robots use two kinds of sensors during their operation: *internal* or proprioceptive *sensors* and *external* or exteroceptive *sensors*. A third category of sensors is used for testing and *performance evaluation* purposes.

Internal sensors, like encoders, resolvers, tachometers, accelerometers, etc. are traditionally used for closing the positioning loops of the different robot axes. In the framework of the SACODY project, additional encoders have been used on the three main robot axes of the KUKA robot, to measure the joint flexibilities. A special sensor system, called *DIOMEDES*, aimed at measuring on-line the robot link deflections, has been developed within the SACODY project, and tested on a breadboard system in a real time control situation. It is described in detail in section 2.

External sensors are used to let the robot interact with the environment and to collect information from that environment. Typical examples are force/torque wrist sensors, tactile sensors, vision systems, etc. These sensors have not been explicity developed within SACODY, although occasionally force/torque wrist sensors have been used, as internal sensors for state feedback, for some demonstrations on flexible robot control, at the ESPRIT Conference in Brussels in November 1989.

Special attention had to be paid within SACODY to the development of suitable *sensor systems to assess the performance* of the developed methods and algorithms for the control of flexible robots. Indeed as one of the main thrusts of SACODY was the development of effective, practically useful algorithms for stable vibration-free control of robots with inherent flexibility, it was absolutely necessary to dispose of accurate means to measure and to record the dynamic trajectories of the robot end effector, for analysis and comparison purposes.

Two systems have been developed and fully tested:

- *RODYM*, a system based on a digitizing tablet, suitable for 2D-trajectories, and
- *LASER-TREK*, a fully fledged 3D-system, based on triangulation, using two laser beam tracking units and a retro-reflector mounted on the end effector.

These systems are described in detail in sections 3.4 and 3.5 respectively.

2. DIOMEDES

2.1. Flexible Link Deflections

The deflections to be measured are the deflections at the tip of each flexible link of a robot. This measurement is crucial because the deflections affect the position and orientation of the robot end-effector [1].

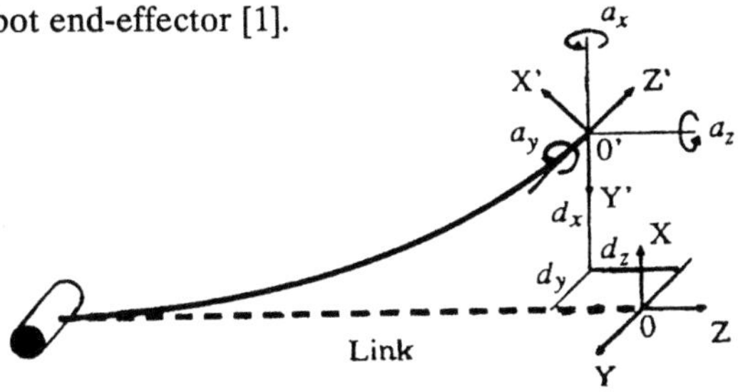

Figure 1: Structural deflections of a robot link

Figure 1 defines the deflections to be measured. Frame $\{X_0, Y_0, Z_0\}$ represents the end of the straight link, before deflection. The z-axis is in the direction of the neutral axis of the link. The frame $\{X_{0'}, Y_{0'}, Z_{0'}\}$ respresents the end of the link after deflection. The description of the deformed link frame with respect to the undeformed link frame is given by the homogeneous transform ${}^0A_{0'}$ [2], defined by:

$$
{}^0\mathbf{A}_{0'} = \begin{bmatrix} c_z c_y & -s_z c_x + c_z s_y s_x & s_z s_x + c_z s_y c_x & d_x \\ s_z c_y & c_z c_x + s_z s_y s_x & -c_z s_x + s_z s_y c_x & d_y \\ -s_y & c_y s_x & c_y c_x & d_z \\ 0 & 0 & 0 & 1 \end{bmatrix}, \qquad (1)
$$

where $c_i = \cos a_i$ and $s_i = \sin a_i$ for $i = x, y, z$. d_x, d_y, d_z, a_x, a_y and a_z are the deflections to be measured. The deflections are the translations along and the rotations about the proper axes of the undeformed link frame (*0-xyz*).

For very small deflections, the trigonometrical functions are linearized, for $i = x, y, z$,

$$
c_i = 1, \quad s_i = a_i
$$

The homogeneous transform ${}^0A_{0'}$ becomes

$$
{}^0\mathbf{A}_{0'} = \begin{bmatrix} 1 & -a_z & a_y & d_x \\ a_z & 1 & -a_x & d_y \\ -a_y & a_x & 1 & d_z \\ 0 & 0 & 0 & 1 \end{bmatrix}. \qquad (2)
$$

2.2. Measurement Set-up

The sensor system to measure the structural deflections d_x, d_y, a_x, a_y and a_z is shown in figure 2. The elongation of the link, d_z can not be measured. Since d_z is usually very small compared to the other deflections, neglecting d_z causes no real error. So, the system is able to measure the global deformation of a robot link.

The system consists of three semiconductor laser diodes mounted at one end of the link and of these PSDs of a planar type, one mounted in the focal plane of a lens, at the other end. Since the laser beams go straight in spite of the deflections of the link, the positions of the spots that the laser diodes make on the PSDs contain all information regarding the structural deflections of the link.

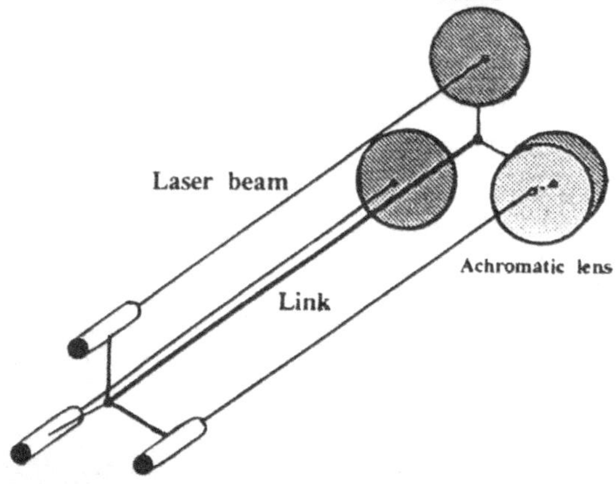

Figure 2: Laser diode optical sensor system to measure the structural deflections of a flexible link

The problem of determining five structural deflections out of six co-ordinates can be divided into two subproblems. The structural deflections d_x, d_y and a_z are calculated by combining the measurements of the PSDs *without a lens*. The influence of the structural deflections a_x and a_y on the measurements should first be eliminated. The structural deflections a_x and a_y are calculated out of the measurements of the PSD *mounted in the focal plane of a lens*. As the PSD is mounted in the focal plane of the lens, the structural deflections d_x and d_y don't affect the position of the laser spot on the PSD. The influence of the structural torsion, a_z, can be avoided if the structural deflections a_x and a_y are defined with respect to the *rotated* link frame.

To determine the relationship between the deflections and the co-ordinates of the laser spots on the PSDs, *co-ordinates frame, homogeneous co-ordinates and homogeneous transformations* are introduced [2]. Figure 3 shows the geometrical relation between the link and a PSD. Frame (X_r, Y_r) represents the link frame (x_{ci}, y_{ci}) the PSD. The frame (x_{ci}, y_{ci}) will be called the *corrected sensor* frame. Position 1 is the position of the sensor frame for a undeformed link, with the origin of the frame coinciding with position of the laser spot on the PSD. The description of the sensor frame with respect to the link frame is given by the homogeneous transform ${}^r\!A_1$, defined by:

$$ {}^r\mathbf{A}_1 = \left[\begin{array}{c|c} {}^r\mathbf{R}_1 & {}^r\mathbf{P}_1 \\ \hline 0\ \ 0\ \ 0 & 1 \end{array}\right] = \left[\begin{array}{ccc|c} 1 & 0 & 0 & S_ic_i \\ 0 & 1 & 0 & S_is_i \\ 0 & 0 & 1 & 0 \\ \hline 0 & 0 & 0 & 1 \end{array}\right], \qquad (3) $$

with ${}^{r}\boldsymbol{R}_{1}$ the orientation matrix and ${}^{r}\boldsymbol{P}_{1}$ the position vector. S_i is the distance between the two frames, α_i is the angle between the position vector to the origin of the sensor frame and the X-axis of the link frame, $c_i = \cos \alpha_i$ and $s_i = \sin \alpha_i$.

Due to the structural deflections d_x , d_y and a_z the PSD moves from position 1 to position 2 and the laser spot remains at position 1. This is only true for the PSDs *without a lens*.

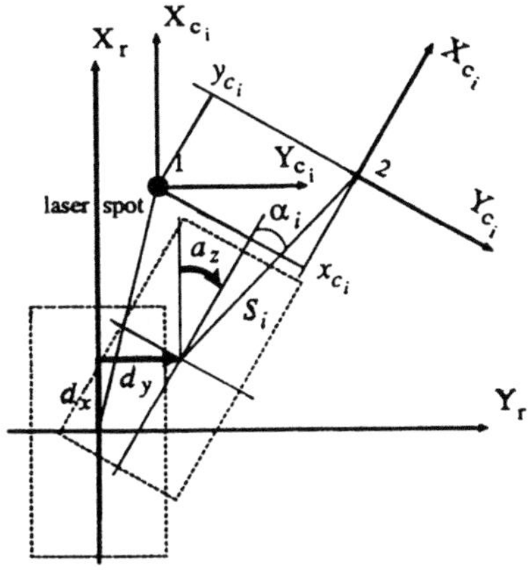

Figure 3: Geometrical relation between the link and a PSD

The homogeneous transform for ${}^{r}\boldsymbol{A}_2$ is given by

$$ {}^{r}\boldsymbol{A}_2 = \boldsymbol{A}\,{}^{r}\boldsymbol{A}_1, \qquad (4) $$

with $\boldsymbol{A}$ the transformation matrix due to the structural deflections, defined by:

$$ \boldsymbol{A} = Translation(X_r, d_x)\ Translation(Y_r, d_y)\ Rotation(Z_r, a_z) $$

or

$$ \begin{aligned} \mathbf{A} &= \begin{bmatrix} 1 & 0 & 0 & d_x \\ 0 & 1 & 0 & 0 \\ 0 & 0 & 1 & 0 \\ 0 & 0 & 0 & 1 \end{bmatrix} \begin{bmatrix} 1 & 0 & 0 & 0 \\ 0 & 1 & 0 & d_y \\ 0 & 0 & 1 & 0 \\ 0 & 0 & 0 & 1 \end{bmatrix} \begin{bmatrix} c_z & -s_z & 0 & 0 \\ s_z & c_z & 0 & 0 \\ 0 & 0 & 1 & 0 \\ 0 & 0 & 0 & 1 \end{bmatrix}, \\ &= \begin{bmatrix} c_z & -s_z & 0 & d_x \\ s_z & c_z & 0 & d_y \\ 0 & 0 & 1 & 0 \\ 0 & 0 & 0 & 1 \end{bmatrix}. \end{aligned} \qquad (5) $$

The co-ordinates of the laser spot in the sensor frame are given by the position vector ${}^{2}\boldsymbol{P}_1$, part of the homogeneous transform ${}^{2}\boldsymbol{A}_1$.

${}^{2}\boldsymbol{A}_1$ is defined by

$$ {}^{r}\boldsymbol{A}_2\,{}^{2}\boldsymbol{A}_1 = {}^{r}\boldsymbol{A}_1, $$

$$ {}^{2}\boldsymbol{A}_1 = {}^{r}\boldsymbol{A}_2^{-1}\,{}^{r}\boldsymbol{A}_1. $$

Taking account of equation 4 yields:

$$^{2}A_{1} = (A\,{}^{r}A_{j})^{-1}\,{}^{r}A_{1},$$

$$^{2}A_{1} = {}^{r}A_{1}^{-1}A^{-1}\,{}^{r}A_{1}. \tag{6}$$

Using the properties of homogeneous transforms and denoting M^t the transpose of a matrix M,

$$^{j}\mathbf{A}_{i}^{-1} = \begin{bmatrix} & ^{j}\mathbf{R}_{i}^{-1} & & ^{j}\mathbf{P}_{i}^{-1} \\ 0 & 0 & 0 & 1 \end{bmatrix}, \tag{7}$$

$$^{j}R_{i}^{-1} = {}^{j}R_{i}^{t}, \tag{8}$$
$$^{j}P_{i}^{-1} = -{}^{j}R_{i}^{t}\,{}^{j}P_{i} \tag{9}$$

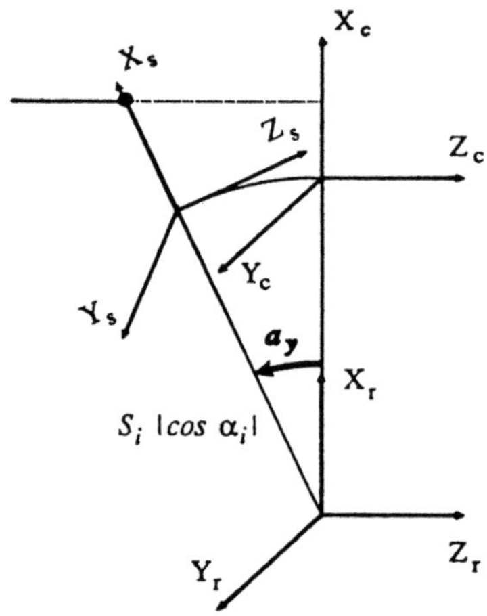

Figure 4: Influence of the bending angle ay on the co-ordinates of the laser spot

The co-ordinates of the laser spot in the sensor frame, given by $^{2}P_{1}$, are calculated as follows

$$^{2}P_{1} = -{}^{r}P_{1} - R^{t}\,(P - P_{1})$$

$$= \begin{bmatrix} -S_i c_i - c_z(d_x - S_i c_i) - s_z(d_y - S_i s_i) \\ -S_i s_i + s_z(d_x - S_i c_i) - c_z(d_y - S_i s_i) \\ 0 \end{bmatrix}.$$

$$x_{ci} = -S_i c_i - c_z\,(d_x - S_i c_i) - s_z\,(d_y - S_i s_i) \tag{10}$$
$$y_{ci} = -S_i s_i + s_z\,(d_x - S_i c_i) - c_z\,(d_y - S_i s_i) \tag{11}$$

The co-ordinates x_{ci} and y_{ci} are the corrected co-ordinates of the laser spot in the sensor frame. Figure 4 shows the influence of the structural deflection ay on the co-ordinate x_{ci}. Equation 12 gives the relation between the corrected co-ordinate x_{ci} and the measured co-ordinate x_{si}.

$$x_{ci} = (x_{si} + l_{xi})\,c_y - l_{xi} \tag{12}$$

The corrected co-ordinate y_{ci} can be calculated as follows:

$$y_{ci} = (y_{si} + l_{yi})\, c_x - l_{yi} \qquad (13)$$

The distances l_{xi} and l_{yi} are defined by

$$l_{xi} = S_i \mid c_i \mid \qquad (14)$$

$$l_{yi} = S_i \mid s_i \mid \qquad (15)$$

where $|.|$ means the absolute value of ..

For the PSD equipped with a lens, figure 5 shows the relation between the structural deflections a_x and a_y , and the position of the laser spot on the PSD. The influence of the torsion, a_x , can be avoided if the structural deflections a_x and a_y are defined with respect to the rotated link frame.

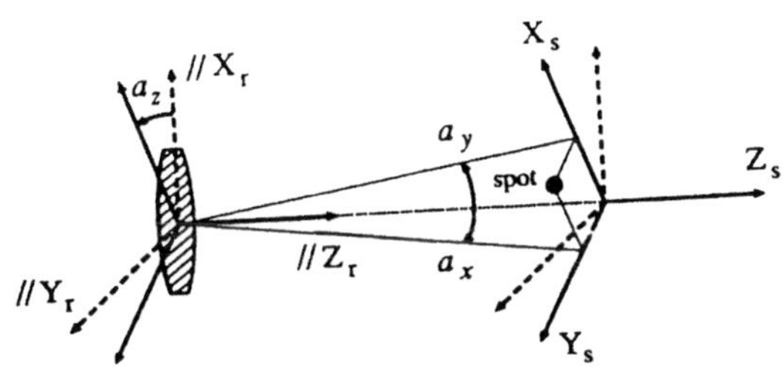

Figure 5: Relation between the structural bending and the position of the laser spot on the PSD mounted in the focal plane of a lens

The co-ordinates of the laser spot in the sensor frame are given by:

$$x_{si} = -f \tan a_y , \qquad (16)$$

$$y_{si} = f \tan a_x , \qquad (17)$$

with f the distance between the lens and the PSD. For an optimally constructed lens-PSD system, f equals the focal length of the lens. In reality, the measuring frame does not always coincide with the assumed sensor frame. The co-ordinates of the laser spot are measured in the measuring frame. The structural deflections are calculated out of the sensor frame co-ordinates. Figure 6 shows the geometrical relation between the measuring frame and the sensor frame expressed by equation 18. The homogeneous transformation ${}^{si}A_{mi}$ corrects the measured co-ordinates.

$$\begin{bmatrix} x_{s_i} \\ y_{s_i} \\ 0 \\ 1 \end{bmatrix} = \underbrace{\begin{bmatrix} \cos\alpha_{0_i} & -\sin\alpha_{0_i} & 0 & x_{0_i} \\ \sin\alpha_{0_i} & \cos\alpha_{0_i} & 0 & y_{0_i} \\ 0 & 0 & 1 & 0 \\ 0 & 0 & 0 & 1 \end{bmatrix}}_{{}^{s_i}A_{m_i}} \begin{bmatrix} x_{m_i} \\ y_{m_i} \\ 0 \\ 1 \end{bmatrix}, \qquad (18)$$

where x_{0i} and y_{0i} are the co-ordinates in the sensor frame of the origin of the measuring frame. The structural angle α_{0i} is measured positively clockwise.

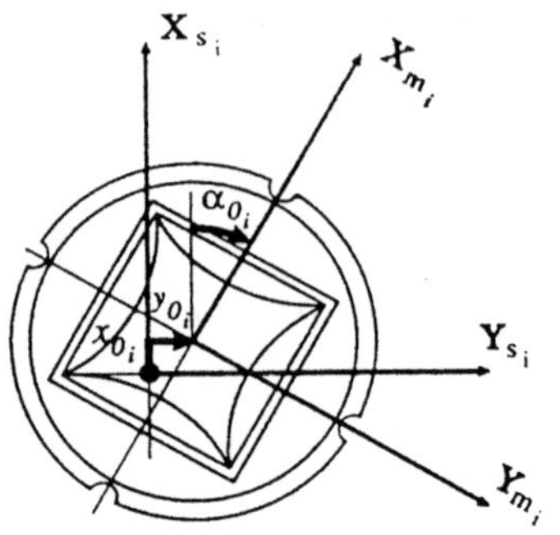

Figure 6: Geometrical relation between the measuring frame and the sensor frame

2.3. Determination of Deflections

2.3.1. Calculation of Deflections

Assume, the PSDs without a lens get the subscript 1 and 2, and the PSD equipped with a lens the subscript 3.

Solving Equations 10 and 11 for the structural deflections d_x and d_y yields

$$d_x = S_i c_i - c_z (x_{ci} + S_i c_i) + s_z (y_{ci} + S_i s_i) \tag{19}$$
$$d_y = S_i s_i - s_z (x_{ci} + S_i c_i) - c_z (y_{ci} + S_i s_i) \tag{20}$$

The structural deflections d_x and d_y are calculated out of equations 19 and 20 if the structural deflection a_z is known. The structural deflections d_x and d_y should be the same for $i = 1$ and $i = 2$.

Therefore:

$$S_1 c_1 - c_z (x_{c1} + S_1 c_1) + s_z (y_{c1} + S_1 s_1) = S_2 c_2 - c_z (x_{c2} + S_2 c_2) + s_z (y_{c2} + S_2 s_2) \tag{21}$$
$$S_1 s_1 - s_z (x_{c1} + S_1 c_1) - c_z (y_{c1} + S_1 s_1) = S_2 s_2 - s_z (x_{c2} + S_2 c_2) - c_z (y_{c2} + S_2 s_2) \tag{22}$$

Solving equations 21 and 22 for a_z yields:

$$\Delta = (y_{c2} + S_2 s_2 - y_{c1} - S_1 s_1)^2 - (x_{c2} + S_2 c_2 - x_{c1} - S_1 c_1)^2$$

$$s_z = [(S_2 s_2 - S_1 s_1)(x_{c2} + S_2 c_2 - x_{c1} - S_1 c_1) - (S_2 c_2 - S_1 c_1)(y_{c2} + S_2 s_2 - y_{c1} - S_1 s_1)] / \Delta \tag{23}$$

$$c_z = [(S_2 c_2 - S_1 c_1)(x_{c2} + S_2 c_2 - x_{c1} - S_1 c_1) + (S_2 s_2 - S_1 s_1)(y_{c2} + S_2 s_2 - y_{c1} - S_1 s_1)] / \Delta \tag{24}$$

$$a_z = \arctan(s_z / c_z) \tag{25}$$

The structural deflections a_x and a_y are calculated out of Equations 16 and 17, with respect to the rotated link frame.

$$a_x = arctan(y_{s3}/f) \tag{26}$$
$$a_y = arctan(-x_{s3}/f) \tag{27}$$

2.3.2. Calibration of Set-up

To be able to calculate the structural deflections, the measurement set-up is first calibrated. Within the equations of the previous subsection, some system parameters appear beside the co-ordinates of the laser spot on the PSD.

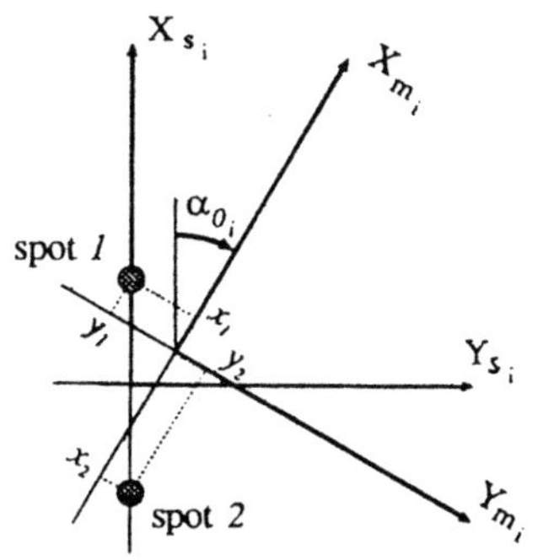

Figure 7: Experiment measuring the angle between the measuring frame and the sensor frame

1. The homogeneous transform of the measuring frame with respect to the sensor frame ${}^{si}A_{mi}$, for $i = 1, 2, 3$.
2. The distance between the sensor frame and the link frame S_i, and the angle between the position vector to the origin of the sensor frame and the X-axis of the link frame α_i, for $i = 1,2$.
3. The distance between the lens and the PSD f, for $i = 3$.

The homogeneous transform ${}^{si}A_{mi}$ is needed to correct the measured co-ordinates. Figure 7 shows the experiment to determine α_{0i}. Position 1 is the position of the laser spot on the PSD for the unloaded link, position 2 is for the link loaded so the link translates only in the X-axis direction of the sensor frame, equals the X-axis direction of the link frame.

$$\alpha_{0_i} = -\frac{y_2 - y_1}{x_2 - x_1}, \tag{28}$$

x_{0i} and y_{0i} are determined by measuring the position of the laser spot on the PSD for the unloaded link. Transforming the co-ordinates from the measuring frame to the sensor frame yields x_{0i} and y_{0i}.

$$x_{0i} = -x_i \cos\alpha_{0i} + y_i \sin\alpha_{0i} \tag{29}$$
$$y_{0i} = -x_i \sin\alpha_{0i} - y_i \cos\alpha_{0i} \tag{30}$$

The distance between the sensor frame and the link frame S_i, and the angle between the position vector to the origin of the sensor frame and the X-axis of the link frame α_i are determined out of equations 10 and 11, for $i = 1,2$. For a pure torsion of the link, with known torsion angle a_z, S_i is solved out of equations 10 and 11.

$$x_{ci} = - S_i c_i + c_z S_i c_i + s_z S_i s_i,$$
$$y_{ci} = - S_i s_i - s_z S_i c_i + c_z S_i s_i,$$
$$x_{ci}^2 + y_{ci}^2 = 2 S_i^2 (1 - c_z) = 4 S_i^2 s_{z/2}^2$$

$$S_i = \frac{\sqrt{x_{c_i}^2 + y_{c_i}^2}}{2 s_{z/2}}. \qquad (31)$$

Solving equations 10 and 11 for α_i yields

$$-c_i + c_z c_i + s_z s_i = x_{ci} / S_i,$$
$$-s_i - s_z c_i + c_z s_i = y_{ci} / S_i,$$
$$(c_z - 1) c_i + s_z s_i = x_{ci} / S_i,$$
$$- s_z c_i + (c_z - 1) s_i = y_{ci} / S_i.$$

$$s_i = \frac{y_{c_i}(c_z - 1) + x_{c_i} s_z}{2 S_i (1 - c_z)}, \qquad (32)$$

$$c_i = \frac{x_{c_i}(c_z - 1) - y_{c_i} s_z}{2 S_i (1 - c_z)}, \qquad (33)$$

$$\alpha_i = \arctan \frac{s_i}{c_i}. \qquad (34)$$

For pure torsion, $a_x = 0$ and $a_y = 0$. So, $(x_{ci}, y_{ci}) = (x_{si}, y_{si})$.

The distance between the lens and the sensor f is determined out of equations 16 and 17, for $i = 3$. For a bending of the link, with known bending angles a_x and a_y, f is solved out of equations 16 and 17.

$$x_{s3}^2 + y_{s3}^2 = f^2 (\tan^2 a_x + \tan^2 a_y),$$

$$f = \frac{\sqrt{x_{s_3}^2 + y_{s_3}^2}}{\sqrt{\tan^2 a_y + \tan^2 a_x}}. \qquad (35)$$

The sum and the square are needed to correct the influence of a_z on the co-ordinates.

2.4. Experiments

2.4.1.- Experimental Set-Up

The purpose of the experimental set-up is to demonstrate the feasibility of the sensor system to determine in *real time* the spatial structural deflections of a flexible link. Figure 8 shows the experimental set-up of *DIOMEDES*. The link is a rectangular beam made of aluminium. Three semiconductor laser diodes are mounted at one end of the link. The laser diodes deliver a maximum output power of 1mW and emit light at a wavelength of *790 nm*. The laser diodes have a diameter of *8 mm* and a length of *17 mm*. For two laser diodes, the laser beam is focused over *500 mm*. For the third laser diode, the laser beam is focused over *800 mm*. Three PSDs are mounted at the other end of the link. One PSD is mounted in the focal plane of a lens and has an effective sensitive area of *2 x 2 mm*. The lens is an achromatic lens with a focal length of *56 mm*. The PSDs without a lens have an effective sensitive area of *12 x 12 mm*. The laser diodesand the PSD equipped with a lens are easily adjustable concerning their pointing accuracy. Analog evaluation electronic circuits are developed to measure the co-ordinates of the laser spot on the PSD. The co-ordinates are fed to the calculation algorithm in the microcomputer using a *12-bit* Lab Master DMA board. The microcomputer is a PC-AT, operating at *8 MHz* and equipped with a floating point co-processor. The software allows to calibrate the set-up and to calculate, in real time, the five structural deflections. A graphical simulation routine is available to show continuously the spatial deformation of the end of the flexible link.

Figure 8: Experimental set-up of DIOMEDES

The presented sensor system has a measuring range of *+/- 5 mm*, with a resolution of *3 μm*, for the structural deflections d_x *and* d_y depending on the value of the torsion angle a_z, and a measuring range of *+/- 14 mrad*, equivalent to *+/- 0.8°*, with a resolution of *7 μrad*, for the bending angles a_x and a_y. The time to measure the six co-ordinates and to calculate the five structural deflections is *5 msec*. The bandwith of the sensor system is over *50 Hz*. The system is able to measure static as well as dynamic deformations of flexible links. To demonstrate the *real time* applicability of the sensor system, the sensor system will be used to improve the accuracy of the tracking control of a flexible two-link manipulator.

2.4.2. Preliminary Experimental Results

The aluminium beam is excited by a vertical sinusoidal force with an amplitude of *50 N* and a frequency of *20 Hz* applied to the end of the beam. Figure 9 shows the measured laser spot co-ordinates from the different PSDs. The DC components are the laser spod co-ordinates before the excitation of the beam. Figure 10 shows the structural deflections calculated from the measured laser spot co-ordinates.

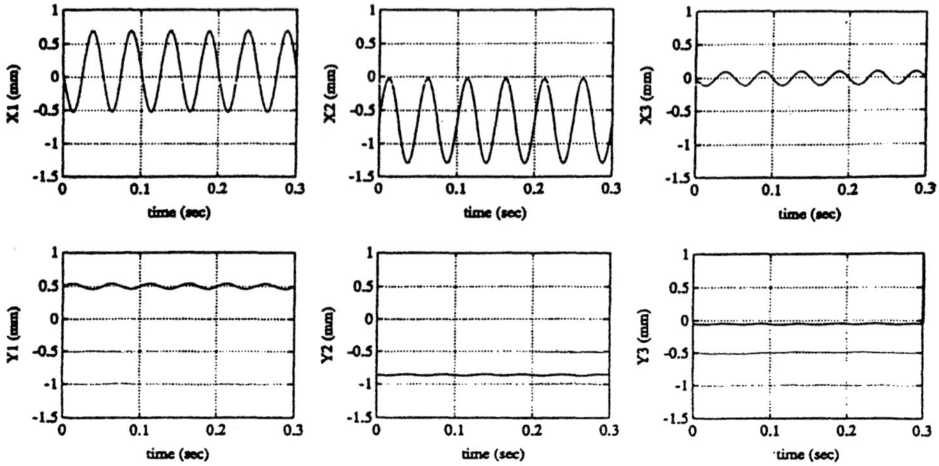

Figure 9: Measured laser spot co-ordinates

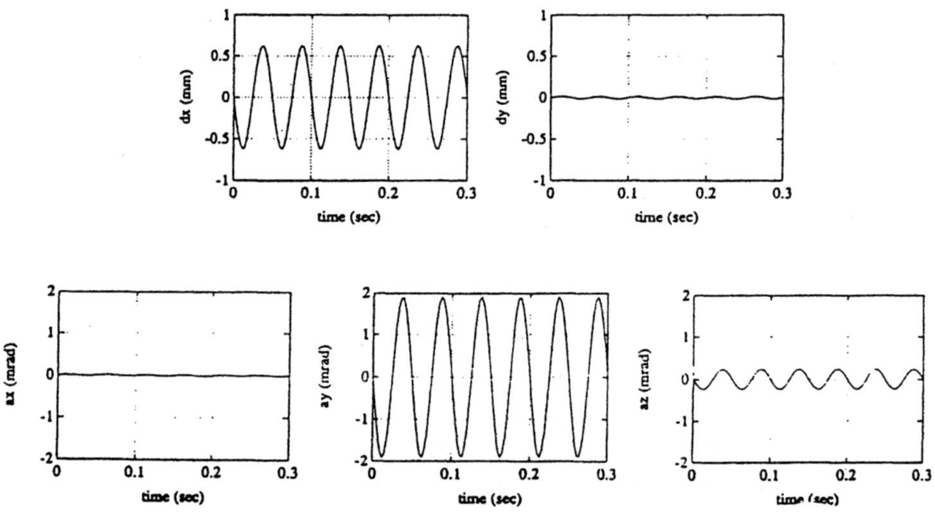

Figure 10: Calculated structural deflections

3. Sensors for Evaluation and Testing of Robot Performance

3.1. Errors Affecting Robot Accuracy

The *position error* of a frame, attached to the robot end effector, is defined as the difference between its actual position and the programmed position. This position error is the summation of the static *position error* or *pose error* and the *dynamic error* or *path error* due to the robot motion.

The pose of the end effector frame is expressed by a *position vector* of end effector frame origin 0 with respect to a reference frame (base frame), and a *rotation matrix* (expressed in Euler angles, Roll-Pitch-Yaw angles, direction cosines, orientation vector, etc.), describing the orientation of the end effector frame with respect to the base frame. Both pose components, position and orientation, can be expressed in one single *transformation matrix* A.

The *static pose* error can consist of the following components:

- digitization error,
- calibration error,
- deterministic kinematic error,
- stochastic kinematic error.

Digitization errors are due to finite encoder resolution on each robot axis and to computer round-off errors. *Calibration errors* occur when, for the robot in home position, the actual joint coordinates differ from the ones assumed by the controller when performing inverse kinematics. *Deterministic kinematic errors* are caused by differences between the actual kinematic parameters (length of links,...) and the ones used in the kinematic model used for inverse kinematic coordinate transformation. Finally, *stochastic kinematic errors* are due to imperfections, like friction or clearance, meaning that, within an encoder resolution, a robot link can assume an arbitrary position, depending on its motion history. The magnitude of this error directly affects the robot repeatability. *Environmental effets*, e.g. thermal gradients are another source of (stochastic or deterministic) kinematic errors.

Dynamic errors affect the path accuracy. A first source of dynamic errors is due to *inertia* and/or *gravity effects*, causing configuration and acceleration dependent link and joint deflections. Second, *dynamic tracking errors* are inherent to servo systems and are the main source of robot dynamic errors.

Many of the discussed static errors can be kept within acceptable limits by a good mechanical design. Consequent application of the mechatronics design methodology enables the compensation of dynamic errors by advanced controller design [3].

3.2. Performance Criteria for Robots

Because robots come in a large variety of types and configurations, standardization of robot specifications is difficult. We need invariant characteristics in order to be able to compare those specifications. The characteristics can be divided into three categories: *general, pose and path characteristics.*

The main *general characteristics* are: *load, velocity, acceleration, positioning time* and *working space.*

The *load carrying capacity* of a robot must be specified in an unambiguous way. Normally the load limitations are given as a function of the distance between the center of gravity of the load and the center of the gripper mounting surface. For eccentric loads, the allowable moments of inertia must be specified.

By velocity is meant the *path velocity* of the center point of the arm/gripper interface surface (endpoint) or of a relevant point associated with the tool function (*tool center point*). Important characteristics are *accuracy, repeatability and fluctuation of the path velocity*. The *path velocity accuracy* is the error in percent between the commanded velocity and the mean value obtained from repeated measurements. *Path velocity repeatibility* is the closeness of agreement of the velocities attained under the same conditions. *Path velocity fluctuation* is the maximum deviation in velocity during one replication for a certain commanded velocity.

Acceleration data can be provided at two levels: acceleration of individual axes, and the *path acceleration* of the endpoint or tool center point. For the user, the path acceleration is more important than the axis acceleration.

The *positioning time* of a robot is the elapsed time needed by the robot to move along a predetermined path, between a stationary start point and a stationary target point. This time thus includes stabilization or settling time. Positioning time is a very important characteristic for applications like spot welding.

The *working space* is the set of poses relative to the reference frame which can be reached by the endpoint. For most robot configurations, it is very difficult to know the working space accurately, especially when orientation of the end effector is to be taken into account.

The accuracy with which a robot can bring and hold a load in a certain pose is probably the most important robot characteristic. Important *pose characteristics* are *pose accuracy* and *pose repeatability*. *One-way pose accuracy* expresses the deviation between a commanded pose and an attained pose, approached in the same direction. *One-way pose repeatability* is the closeness of agreement between the measured pose, for the same commanded pose, after making the robot reach the pose several times from the same direction, under the same quasi-static motion conditions. When the restriction of unidirectional approach is lifted, then we obtain a *multi-way pose accuracy and repeatability*.

Distance accuracy expresses the deviation between the programmed distance and the average of the measured distance, both in position and orientation. *Distance repeatability* is the closeness of agreement between both position and orientation measured distance, for the same programmed distance, after moving over the same distance several times in the same direction.

Stabilization time is the period of time which elapses between the instant when the robot control issues the attained pose signal and the instant at which the damped oscillatory motion of the end or tool center point falls within some specified amplitude band. The *pose overshoot* is the maximum deviation between the commanded pose and the attained pose in the approach phase, counted from the moment the robot issues the attained pose signal. *Drift* is the variation of pose accuracy over a specified time interval.

For continuous path applications, *path characteristics* are more important than pose characteristics. Several researchers have attempted to define relevant general path characteristics. In this paper, the ISO approach has been adopted. ISO defines path accuracy features similar to those defined for pose characteristics. *Path accuracy* is the maximum deviation from the commanded path obtained from following the same path several times in the same direction. *Path repeatability* expresses the closeness of agreement between the pose of the end effector along the same path, followed several times in the same direction.

Robot behavior at path discontinuities is expressed by *cornering overshoot* and *round-off errors.* For standardization purposes, the commanded path considered consists of two straight lines perpendicular to each other, crossing at the *edge point. Cornering round-off error* is defined as the minimum distance between the edge point pose and the obtained path. *Cornering overshoot* is the maximum deviation from the programmed path after the robot passed the edge point. The *stabilization path length* is defined as the length from the edge point to the point along the second commanded path line, frome where the robot performs again according to the specified path characteristics.

Performance criteria of a robot with respect to path velocity are divided into three categories: *path velocity accuracy, path velocity repeatability and path velocity fluctuation.* The *path velocity accuracy* is defined as the error between the commanded velocity and the mean value of the attained velocitites achieved during n repeat traverses along the standard test path. *Path velocity repeatibility* is a measure of the closeness of agreement of the velocities attained for the same commanded velocity. The *path velocity fluctuation* is the maximum deviation in velocity during one replication with the commanded velocity.

3.3. Standardization Efforts

Considerable effort has been spent over the last two decades in the main industrialized countries, to try to standardize the methods for assessing the performance of industrial robots. This activity has resulted in a number of draft standards or guide lines, the most important of which are the ISO/DIS 9283 Draft International Standard, the VDI 2861 Richtlinien, the ANSI/RIA R15.05-1 American National Standard. These standards are discussed in more detail hereunder and compared with each other.

3.3.1. ANSI/RIA R15.05 Industrial Robot Performance Test Standard [4]

The objective of the ANSI/RIA R15.05 Robot Performance Test Standard is to provide meaningful technical information to help robot users select the best robot for their application. It defines the most important performance criteria and a method for evaluating these criteria. The criteria are static position accuracy, cycle time, overshoot, settling time, compliance and point to point repeatability and path accuracy. Both will appear in the future ANSI/RIA Robot Dynamic Performance Test Standard.

In order to achieve relative comparison of robots, standard test path and conditions are used. A *standard test plane* is chosen, orthogonal to the (1, 1, -1) vector (Figure 11a) of the robot base frame. A *standard test path* is located in this standard test plane. It is composed of a number of U-formed segments (Figure 11b). All data will be taken from measurements of motion in the standard test path.

Standard test conditions define the environment and conditions under which the test should be performed. Ambient temperature, relative humidity and test load are specified.

Performance classes are defined to classify the robot performance when used at rated capacity, to optimize maximum cycle rate, repeatability or other specific criteria important for certain applications. These criteria can be optimized by changing operating parameters (e.g. servo loop feedback gain).

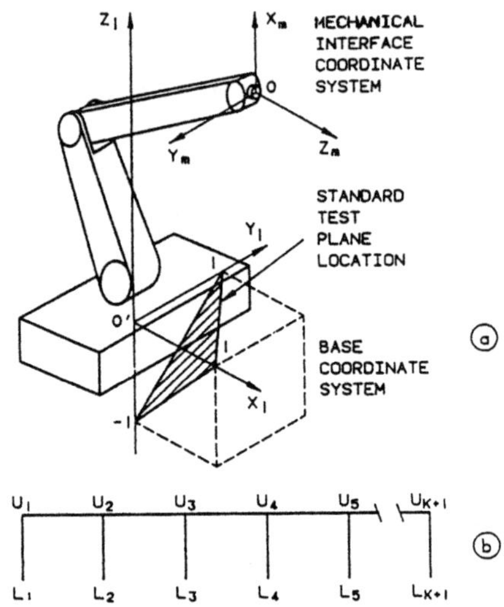

Figure 11: Standardized test plane (a)
and standard test path (b) according to ANSI/RIA R15.05

Different *performance criteria*, defined in a similar way as in section 3.2, are chosen such as to represent the best indication of the overall performance of industrial robots.

The *static position accuracy* is statiscally expressed by its *mean*, *standard deviation* and *maximum measured* (the most positive individual observation in the sample), obtained from several measurement.

Two statistics are used for *static position repeatability*: *warm-up drift* and *operational repeatability*. Warm-up drift is the difference between the first toolpoint placement (cold start) and the placement after stabilization. Operational repeatability is the $+/- 3\sigma$ sphere radius that would bound *99.7 %* of the repeatability errors.

Two figures of merit characterize the *cycle time*: the *segment cycle time* and the *traverse speed*. The *segment cycle* time is a measure of the average time required by the robot to travel through one segment of the standard test path. Each segment motion consits of an upward movement, a horizontal movement, and a downward movement. The *traverse speed* is a measure of the average speed achieved by the robot while traveling from the first point of the standard test path to the last one.

The *overshoot* is defined here as the largest overtravel distance past the target position along the direction of motion after a command to move to a point. The purpose of this overshoot measurement is to quantify the robot's capability to make smooth and accurate stops at target points. Such capability is an important performance parameter for applications which involve a large inertia, high speed or frequent stops. The *settling time* is the period of time required for the robot to reach within a limit (plus and minus the repeatability) from the target point after a move command is executed. The settling time quantifies how quickly a robot can stop at the target point. This is important for applications such as spot welding that require many stops.

The *robot compliance* is the structural characteristic of a robot which allows the robot to deform under external forces, both statically and dynamically. It is induced by relative motion between torque transmission mechanisms at joint and actuators, deflection of linkages, etc... The static compliance is the elastic deflection of a robot under a static force. It acts as a forgiving factor while the robot is interacting with the outside environment. It makes the robot accommodate limited positional errors during assembly operations, react correctly to an uneven force distribution during spot welding operations, etc... On the other hand, machining operations require stiff robots to accurately hold work pieces. Thus, it is important that the compliance of a robot suits the particular application.

Path accuracy figures of merit are not included yet in the present ANSI/RIA R15.05 draft standard.

3.3.2. ISO/DIS 9283 DRAFT INTERNATIONAL STANDARD [5]

The ISO/DIS 9283 Draft International is similar to ANSI/RIA R15.05, but it is more complete.

It provides guidelines on how to measure the following performance characteristis:

- Pose accuracy and repeatability,
- Pose stabilization time,
- Pose overshoot,
- Drift of pose accuracy,
- Path accuracy and repeatability,
- Cornering overshoot and round-off error,
- Path velocity accuracy and repeatability,
- Minimum positioning time,
- Static compliance.

The poses to be tested and the paths to be followed are located in one of four planes, placed inside a cube within the working space (Figure 12). Five pose points (P_1 - P_5 in Figure 12) are located on the diagonals of the selected plane. Test velocities for pose characteristics are *100, 50* and *10 %* of the rated velocity. The test paths, linear or circular, are also defined inside the selected test plane.

In ISO/DIS 9283, the measurement of a pose requires the *measurement of the complete pose*: three positions and three orientations. ANSI/RIA only requires the measurement of the three position components. ISO/DIS 9283 further provides the same tests as ANSI/RIA, except for the *distance accuracy* and *repeatability tests* and the *velocity test*. The *distance accuracy* and repeatability tests quantify the deviations which occur in the distance and orientation between two commanded poses, and the fluctuations in distances and orientations for a series of repeat movements between the two poses. The concept of distance accuracy and repeatability characteristics in an extension of that of pose accuracy and repeatability.

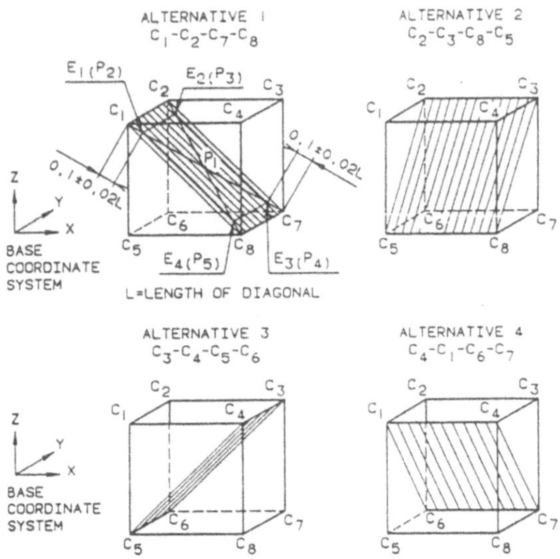

Figure 12: Possible test planes and standardized test paths according to ISO/DIS 9283

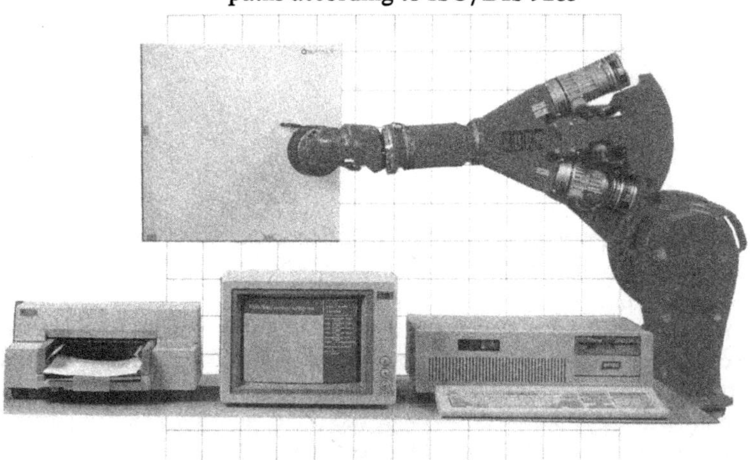

Figure 13: Robot test setup RODYM

3.3.3. VDI 2861 Richtlinien [6]

The VDI 2861 Richtlinien were used as a basis for the development of the ISO/DIS

9283 Draft International Standard, so that both are very similar. The VDI document describes more performance parameters than ISO, e.g. on load characteristics and on geometrical properties. On the other hand, ISO/DIS 9283 puts more emphasis on characteristics relevant to off-line programming, while VDI is mainly restricted to teach-in programming.

3.4. RODYM

3.4.1. System Description

The RODYM system consists of a digitizing tablet, interfaced to a personal computer by an IEEE-488 bus, and a software package for data acquisition and analysis. Figure 13 shows the overall system set-up. *X, Y* coordinates of a measurement probe relative to the surface of the tablet are acquired by a personal computer at a rate of *200 Hz*. To perform tests, this probe is fixed at the end point of the robot to be tested. The tablet is positioned in such a way that the programmed robot movements are in the measurement plane of the board.

The robot base frame *B* and the measurement frame *T* of the digitizer are not coincident. A least squares orthogonal axis transformation is implemented in order to make frame *B* coincident with frame *T*. Bias on the acquired error vectors in eliminated by this transformation. The transformation matrix *M* is calculated by minimizing the sum of all deviations between the commanded points and the acquired points. The achieved (or transformed) points are calculated out of the acquired points by using the matrix *M*.

Different Figures Of Merit (FOM) ranging from *overshoot, settling time* to *static position accuracy* and *path accuracy* are calculated by a menu driven software package (Figure 14). These FOM describe the performance characteristics of the tested robot, as defined in the ANSI/RIA R15.05, ISO/DIS 9283, VID 2861 performance standards for industrial robots. The results of a series of tests are reported in a performance data table (table 1).

The measurement principle of the digitizer board is based on a travelling time measurement of a wave induced by magnetostriction. This wave is propagating along a series of parallel wires and picked up by the probe. The distance of the probe relative to one border of the tablet is related to the wave travelling time. A clock frequency of *50 MHz* is used for this measurement, resulting in a resolution of *0.02 mm*. A calibration technique is developed to improve the accuracy of the digitizer with a factor of four, resulting in a sufficient accuracy *(0.05 mm)* for robot performance analysis.

The sampling frequency of the system is *200 Hz*. This enables to perform dynamic measurements of movements, with a bandwidth up to *20 Hz*. An extension to modal analysis of robots with the proposed sensor is subject to future research.

The surface of the digitizer tablet can range up to *1800 x 1200 mm*. Two modes of acquisition are provided: point to point mode and stream mode. In the first mode a point is acquired each time a command is sent to the digitizer. In the second mode, coordinates are acquired continuously at the maximum sampling frequency. This mode is used for dynamic measurements.

The position of the probe can be acquired at a distance of up to *20 mm* orthogonal to the surface. When the probe is leaving this region, a signal is sent to the personal computer. This feature is used to synchronize RODYM and the robot under test.

RODYM aslo detects if the robot has reached a commanded point, within a repeatability zone.

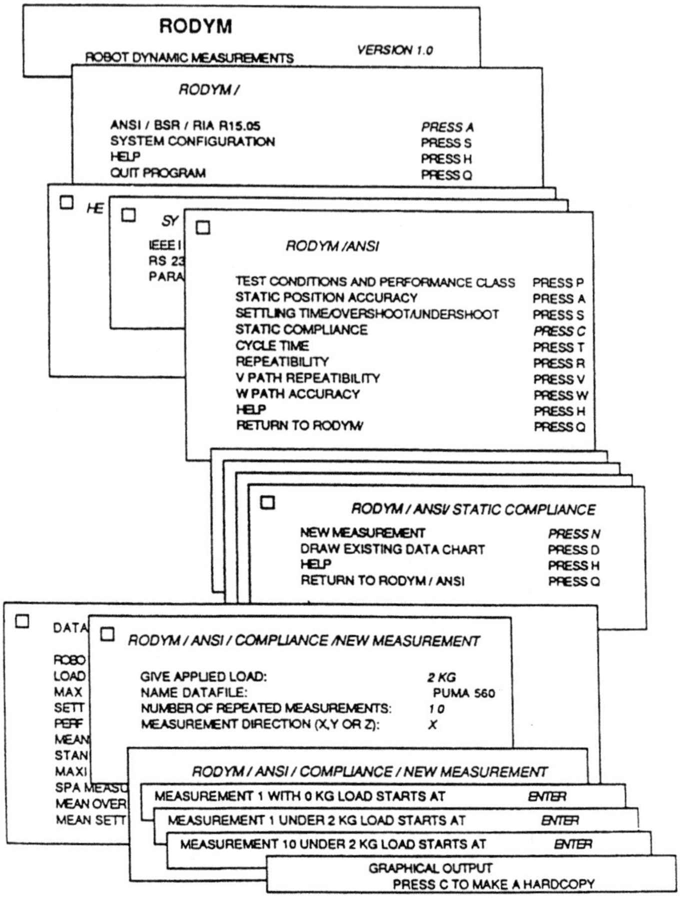

Figure 14: Menu driven robot analysis software

3.4.2. System Error Definition And Calibration

The resolution of the digitizing tablet is *0.02 mm*. The absolute accuracy as provided by the manufacturer is *0.2 mm*. The sampling rate is *200 Hz*. The time delay between acquisition of *X* and *Y* coordinates is *1 ms*. This is causing an error during dynamic measurements; the *Y* coordinate does not correspond to the same point as the *X* coordinate. This error is called the *velocity error* of the tablet. At a speed of *1 m/s*, this error can measure up to *1 mm*. The *static error* of the tablet is composed of three different errors: a *systematic error*, a *statistical error* and a *quantization error*. The *systematic error* is caused by non orthogonality of the measurement grid and deformations in the wires of the tablet. The *statistical error* is generated by noise in the circuit. The *quantization error* is caused by the digitalization process and is equal to half of the value of the least significant bit. Temperature variations affect the clock frequency of the digitizer and the dimensions of the measurement grid.

robot type: Unimation Puma 560	mean segment cycle time: 1.786s
load category: 2kg	traverse speed: 218.4mm/s
maximum velocity: 1m/s	deviation from stp: 10.326mm
settling time: 100ms	repeatability: 0.083mm
performance class: standard	operational 3 σ: 0.155mm
mean static position accuracy: 1.734mm	warm-up drift: 0.160mm
standard deviation spa: 0.934mm	warm-up time: 26min
maximum measured: 2.970mm	mean static X compliance: 0.0252mm/N
spa measurement dwell: 0s	mean static Y compliance: 0.0383mm/N
mean overshoot: 0.549mm	mean static Z compliance: 0.0317mm/N
mean settling time: 0.300s	

Table 1: Performance data chart for a Puma 560

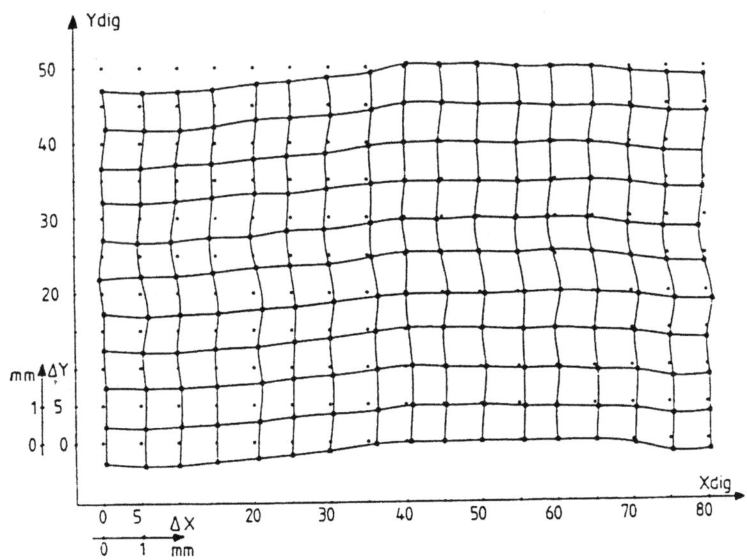

Figure 15: Surface error mapping of digitizer

In order to compensate the *velocity error*, new values Y^* of the Y coordinates are calculated by shifting the latter over a time period of *1 ms*, using a quadratic interpolation. The achieved path can be reconstructes by using the acquired X coordinates and the corrected Y coordinates.

The *static position error* of the tablet is compensated by calibrating both X and Y axes of the digitizer. This calibration is performed with a coordinate measurement machine (CMM). The CMM is programmed to stop at 300 consecutive points along the X axis of the digitizer, with the probe of the digitizing board mounted on the CMMs endpoint. The distance between two consecutive points is *1 mm*. The coordinates are acquired by the digitizing tablet and stored in a vector array X_{corr}. Then an array of points orthogonal to the first one is generated by the CMM to acquire the coordinates of the Y-axis. They are stored in Y_{corr}. These two vector arrays are representing the static errors of both axes of the digitizing board relative to the axes of the CMM.

After mapping a part of the surface *(100 x 100 mm)* (figure 15), it turns out that the measured axis error vectors are transposable over the whole measurement plane. This can be concluded after detecting that the shape of the curve on the left side is the same as the shape on the right side on figure 15. This means that a compensation can be performed by using only information stored in the error vector arrays X_{corr} and Y_{corr}.

3.4.3. Calibration Results With CMM

		without compensation		with compensation	
mean spa (mm)		0.191		*0.051*	
standard deviation (mm)		0.062		*0.018*	
maximum measured spa (mm)		0.283		*0.071*	
commanded X	commanded Y	achieved X	achieved Y	achieved X	achieved Y
30.00	30.00	30.18	30.22	30.04	29.96
130.00	30.00	129.94	30.22	129.96	30.05
230.00	30.00	229.93	30.17	229.96	29.99
230.00	130.00	229.86	129.84	229.98	130.00
230.00	230.00	229.94	229.92	230.03	230.04
130.00	230.00	129.90	229.81	129.98	229.96
30.00	230.00	30.10	229.97	30.02	230.03
30.00	130.00	30.13	129.85	30.06	129.97

Table 2: Static positioning accuracy test on the digitizing tablet

Both calibration vectors where acquired using a Cordax CMM. The accuracy of the board is improved with a factor of four by using the compensation method as described above.

The *static position accuracy* (SPA) test was performed by acquiring eight points at the border of the tablet. *Mean, standard deviation* and *maximum measured* SPA are calculated, using the achieved and the commanded points; the acquired points are transformed to the achieved points by a least square orthogonal axis transformation *(M)*. In table 2, the results of this test are shown with and without compensation algorithm. An improvement of a factor of four on the accuracy figures is reported.

The *path repeatability* test was performed by programming the CMM to travel twice through the same path. The path repeatability is calculated by using a sliding window technique: the distance from a point *P1* on curve A to a series of points in a sliding window on curve B is calculated. If a minimum occurs in this window, then this minimum is compared to the present path repeatability figure. If it is larger, the value of the path repeatability is actualized with this value. If the minimum occurs at one of the endpoints of the window, the window is slided towards this endpoint, until a new minimum can be calculated. The graphical output (figure 16) shows clockwise four windows; the path window, the info window, the data window and the graphics window. The path window indicates the present location in the program. The info window indicates the next step to perform. The data window shows the calculated FOM, in this case the path repeatability. The data file is indicated and the coordinates from the starting point. The system triggers a stream mode measurement when the probe leaves a repeatability zone around this point. The CMM and the measurement system are synchronized in this way. The graphical window shows clockwise the first acquired line, the deviations from both curves in the *X* direction *(DELTA X)*, the second curve, and the deviations form both curves in the *Y* direction *(DELTA Y)*. The calculated FOM are stored in the dat chart (table 2) under the name *deviation from stp*. Either path repeatability or path accuracy can be stored unde this figure. In the situation of figure 16, the obtained path repeatability for the CMM was *0.956 mm*. As we may assume that a CMM is much more accurate, this figure can be considered as the error associated with the digitizing tablet. The *repeatability* of the board was tested by programming the CMM to point a thousand times of a specific location on the board. The *repeatability* of the board is *0.05 mm*.

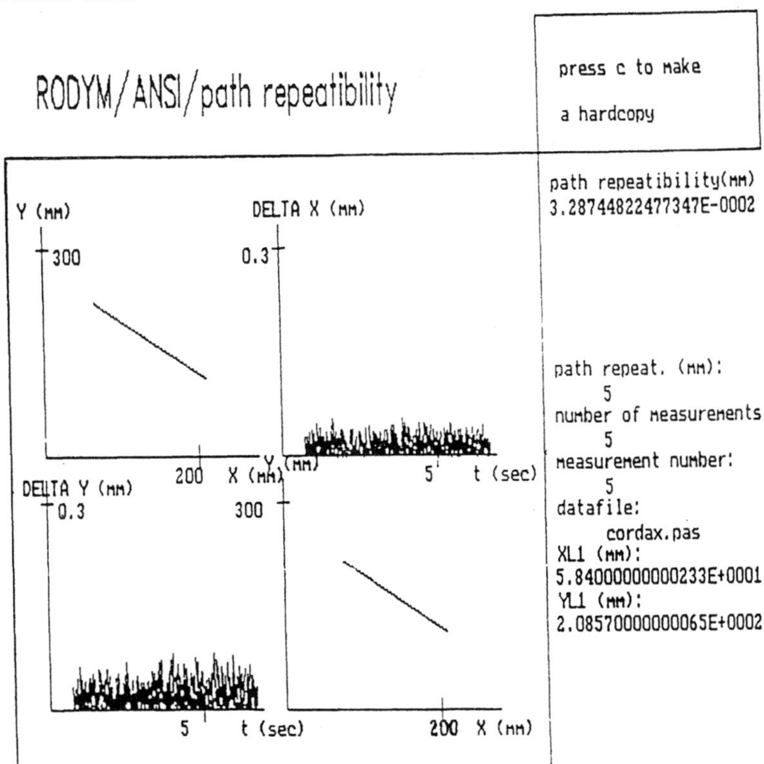

Figure 16: Path repeatability test results

3.4.4. An Example: Measuring a Puma 560 Robot

According to the ANSI/RIA R15.05 performance standard, a number of figures have to be reported. A payload of *2 kg* is mounted on the end effetor, according to the standard performance class. The repeatability test is performed with a cold start. The other tests are done with the robot at operational temperature. The measurement surface of the digitizing board used for these tests is *300 x 300 mm*. With this surface, only a part of the standard test path (STP) can be covered. The data acquisition rate for the dynamic measurements is *200 Hz*. All tests are performed in the Standard Test Path.

Table 1 shows the data chart of the Puma 560. The first five topics are generated by the user. The settling time is provided by the robot manufacturer. The first test figure on the data chart is the *static position accuracy*. Three figures have to be reported: *mean static position accuracy, standard deviation* and *maximum measured static position accuracy.* To perform this test, the tablet was positioned in the STP. The robot is then commanded to stop at eight different points. These eight commanded points are also put into the test program. At each stop, the program acquires the co-ordinates of the commanded point. After the last acquisition, the program calculates the least squares axis transformation matrix M. This matrix is then used to calculate the achieved points. Figure 17 shows the graphical output for this test. Acquired, transformed (or achieved) and commanded points are indicated. After a least square orthogonal axis transformation, the achieved points are almost coincident with commanded points. The *mean* SPA is *1.68 mm*, the *standard deviation* is *0.82 mm*, and the *maximum measured* is *2.75 mm*. The second test is the *overshoot* and *settling-time* test. The tablet is positioned with the X axis parallel to the first vertice of the STP. *Mean overshoot* and *mean settling* time are reported. For the PUMA 560, the *mean overshoot* is *0.59 mm* and the *settling-time* is *0.3 s*. The *static compliance* is calculated by acquiring the endpoint position both under zero and maximum load, along the direction of the applied force. The compliance is measured at point L1 of the STP. The results are reported in table 1.

3.5. LASER-TREK

3.5.1. Operational principle

The developed 3D laser tracking system allows the measurement of the linear coordinates of a point moving in a three-dimensional space. To achieve this, a retro-reflector (corner cube or cat's eye) is connected to the object to be tracked (e.g. a robot end-effector). The optical center of the retro-reflector forms the moving target. The measurements are contactless, the only connection between the measuring device and the measuring point being two laser beams, each originating from a deflection unit and pointing towards the optical center of the moving target. The position units, positioned around the moving target, track the target by keeping the laser beams pointing towards it. The position of the target is obtained by triangulation performed on the triangle formed by the crossing laser beams and the connection line between the points on the deflection mirrors where the laser beams leave (or impinge on) the mirrors. The position of the laser beams in space is determined from the two angular positions of each of the two deflection mirrors. The measuring principle is shown in Figure 18.

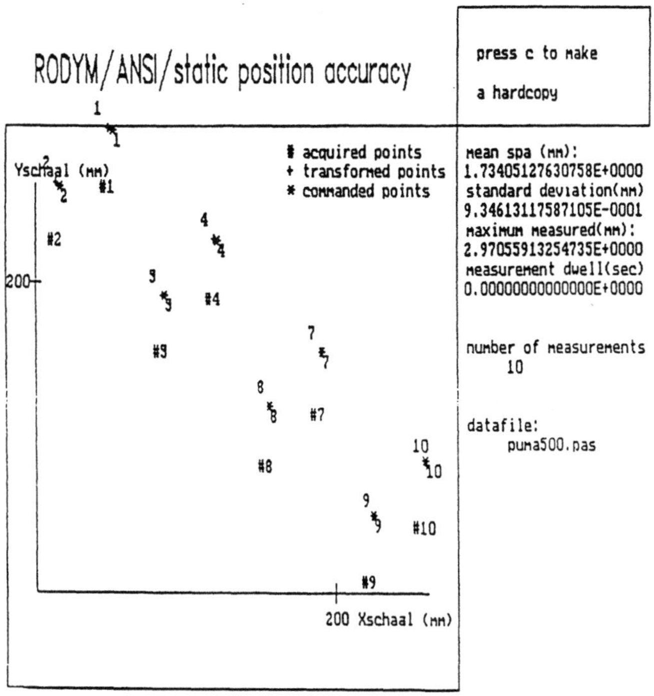

Figure 17: Static position accuracy test of Puma 560

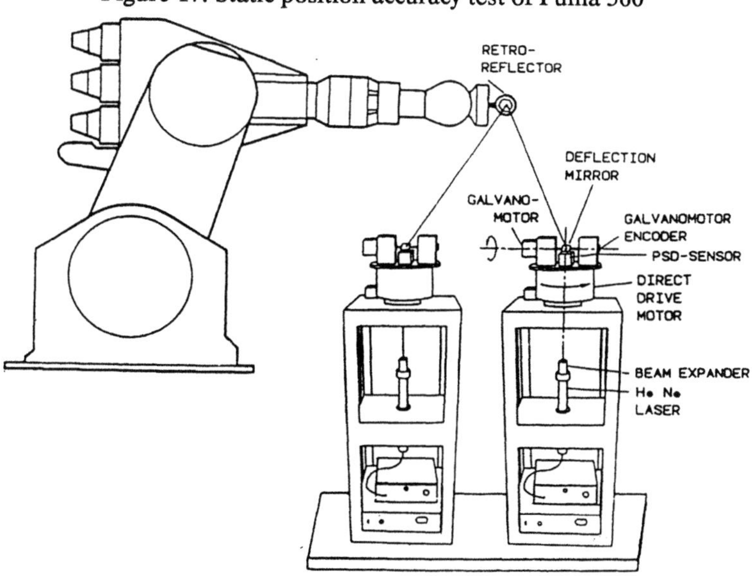

Figure 18: Measuring Principle of LASER-TREK

The measuring principle requires two deflection mirrors, each with two degrees of freedom. Each mirror is mounted on a galvanomotor with horizontal axis, generating a deflection of the laser beam in a vertical plane. An incremental encoder mounted at the free end of the galvanomotor measures the angular displacement Θ_2 of the mirror. This galvanomotor/encoder unit in its turn is mounted on a direct drive motor with a vertical rotation axis. The laser is mounted in the hollow shaft of this motor such that its optical axis coincides with the rotation axis. The direct drive motor deflects the laser beam in a horizontal plane. It has also a built-in high-resolution encoder, measuring the angular displacement Θ_1 of the mirror.

The retro-reflector reflects the laser beam parallel to the incident beam towards the deflection mirror, which makes the reflecting beam impinge on a PSD (Position Sensitive Device)-detector via a beam-splitter. The position of the reflected beam on this detector defines the tracking error. Two signals are generated by this detector, proportional resp. the *x*- and *y*-position of the reflected beam with respect to a coordinate frame, positioned in the center of the PSD. These signals serve as control signals for the mirror positioning motors, which are controlled so as to always direct the laser beams towards the optical center of the retro-reflector. By this the servo-loop is closed, and the two laser beams continuously track the position of the retro-reflector.

3.5.2. THE TRACKING SERVO SYSTEM

The direct drive motor is velocity-controlled and can therefore be modelled by a first-order transfer function which relates the desired and actual angular velocities:

$$G_M(s) = \frac{\alpha(s)}{\alpha_d(s)} = \frac{1}{\tau s + 1}, \tag{36}$$

where τ was identified as $\tau = 1/2\pi f$, with $f = 45\ Hz$.

Figure 19 shows a complete block diagram. After integration, the angular position of the motor is obtained.

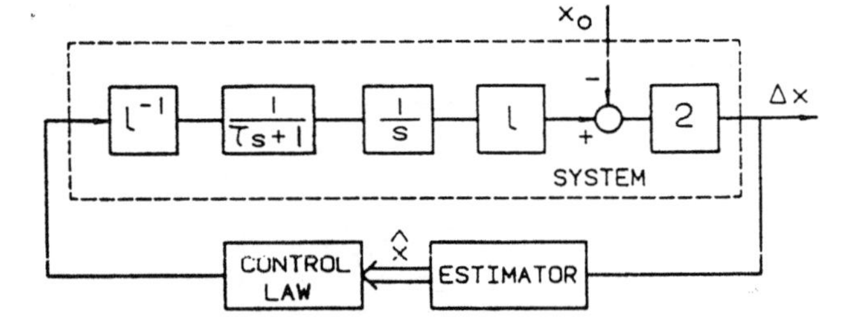

Figure 19: Dynamic model of the tracking system

Multiplication by the distance l between the deflection unit and the retro-reflector yields the absolute position X of the target E in space (Figure 20).

x_0 represents the position of the optical center of the retro-reflector (point O in Figure 20). From Figure 20, it is clear that the difference x-x_0 is amplified by a factor 2 in the retro-reflector. Hence, the PSD signal is:

$$\Delta x = 2\,(x - x_0) \tag{37}$$

The objective is to design a controller $G_c(s)$ which allows a fast tracking of the robot end-effector position x_0. The deflection unit will lose track if the tracking errors exceeds 5 mm (determined by the dimensions of the PSDs).

In order to obtain an equivalent state-space model, an additional factor l^{-1} is considered to be part of the system. By doing so, the obtained control gains are independent of l (see Figure 19). A possible system representation is:

$$\begin{bmatrix} \dot{x}_1 \\ \dot{x}_2 \end{bmatrix} = \begin{bmatrix} 0 & 1 \\ 0 & -\frac{1}{\tau} \end{bmatrix} \begin{bmatrix} x_1 \\ x_2 \end{bmatrix} + \begin{bmatrix} 0 \\ \frac{1}{\tau} \end{bmatrix} u + \begin{bmatrix} -1 \\ 0 \end{bmatrix} \dot{x}_o \tag{38}$$

$$y = \Delta x = \begin{bmatrix} 2 & 0 \end{bmatrix} \begin{bmatrix} x_1 \\ x_2 \end{bmatrix} \tag{39}$$

where $\dot{x}_0$ is considered to be a disturbance input and u is the actual control input. Another possibility is to consider $\dot{x}_0$ as a third state. In this state the system representation becomes:

$$\begin{bmatrix} \dot{x}_1 \\ \dot{x}_2 \\ \dot{x}_3 \end{bmatrix} = \begin{bmatrix} 0 & 1 & -1 \\ 0 & -\frac{1}{\tau} & 0 \\ 0 & 0 & 0 \end{bmatrix} \begin{bmatrix} x_1 \\ x_2 \\ x_3 \end{bmatrix} + \begin{bmatrix} 0 \\ \frac{1}{\tau} \\ 0 \end{bmatrix} u \tag{40}$$

with $x_3 = \dot{x}_0$.

Based on these mathematical models, appropriate controllers $G_c(s)$ have been designed, allowing fast tracking of the robot end-effector positions, at speeds up to *1.5 m/s*.

3.5.3. The Mathematical Model

Based on the four deflection mirror angles, (two for each deflection unit) and four PSD signals, describing the tracking errors originating from the dynamic characteristics of the mirror positioning servos, the position in 3D-space of the tracked points can be determined (Figure 21). The mathematical model consists of a computation scheme to compute the absolute position of the target point p (Figure 22). Therefore, based on the recorded deflection mirror angles and the PSD tracking errors, the spatial positions of the two laser beams are determined with respect to the reference coordinate frame S_h. The intersection point of the two laser beams defines the position of the optical center of the retro-reflector, expressed in the *xyz*-coordinates with respect to the same reference frame. However, as the real system does not perfectly fit the mathematical model, the laser beams computed from the model do not intersect but cross each other. The target position, therefore, is located in the middle of the line perpendicular to the two beams.

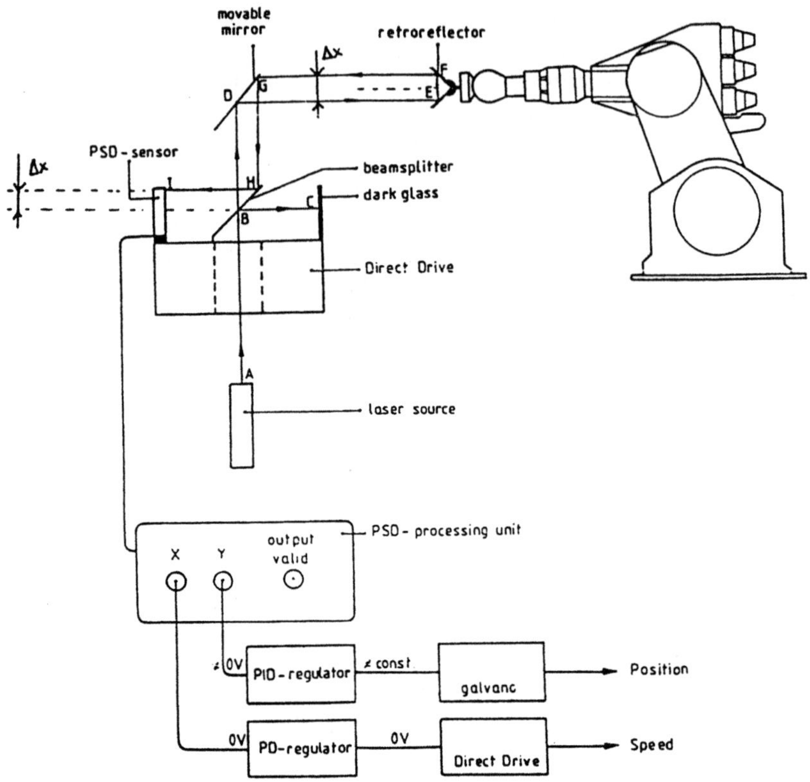

Figure 20: Principle of the tracking error measurement

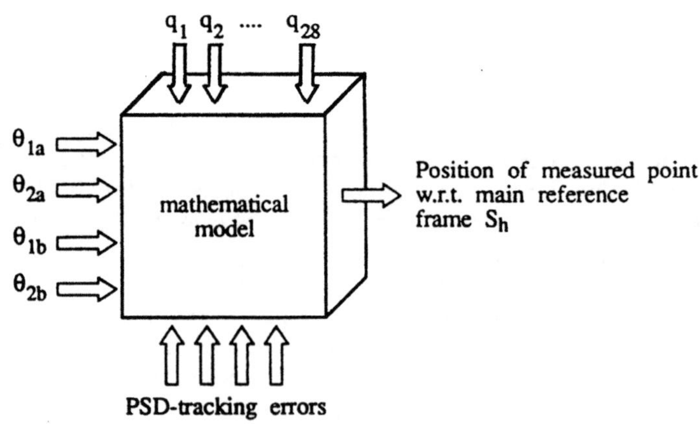

Figure 21: Input-Output model

The coordinates of that point, expressed by vector **p** are given by:

$$\gamma = \frac{(\mathbf{r_b} - \mathbf{r_a}).\mathbf{v}}{\sqrt{\mathbf{v}.\mathbf{v}}}, \tag{41}$$

$$\mathbf{v} = \mathbf{s_a} \wedge \mathbf{s_b}, \tag{42}$$

$$\mathbf{p} = \frac{1}{2}\left\{\mathbf{r_a} + \mathbf{r_b} + \frac{1}{\mathbf{v}.\mathbf{v}}\{[(\mathbf{r_a} - \mathbf{r_b})(\mathbf{v} \wedge \mathbf{s_b})]\,\mathbf{s_a} + [(\mathbf{r_a} - \mathbf{r_b})(\mathbf{v} \wedge \mathbf{r_a})]\,\mathbf{s_b}\}\right\}. \tag{43}$$

where γ is the minimal distance between the two laser beams,
v is the orientation vector along this distance,
indexes *a* and *b* refer to the deflection units *A* and *B* respectively.

In order to take into account the different construction and alignment errors in the system, the ideal mathematical model is extended to include the influence of 28 error parameters q_j listed hereunder (Figure 23).

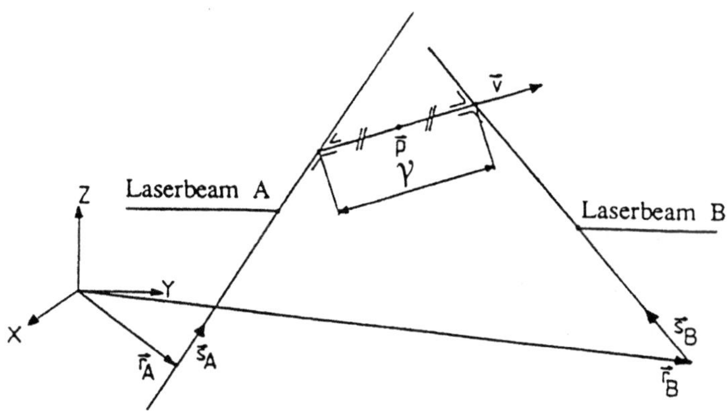

Figure 22: Ideal mathematical model

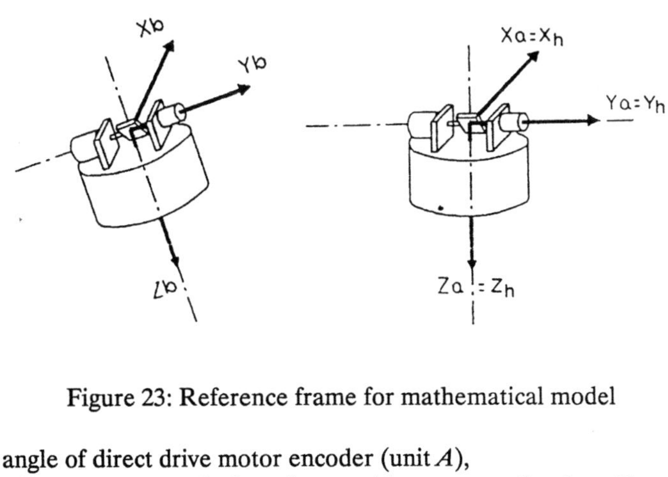

Figure 23: Reference frame for mathematical model

q_1 : offset angle of direct drive motor encoder (unit *A*),
q_2 : linear offset parameter of mirror frame with respect to S_a, along X_a-axis,

q_3 : angular offset of mirror frame S_{sa}, around X_a-axis,
q_4 : offset angle of galvanomotor encoder (unit A),
q_5 : distance between mirror surface and origin of frame S_a,
q_6 : angle between projection of normal to mirror surface in $X_{sa}Y_{sa}$-plane and X_{sa}-axis,
q_7 : angle between projection of normal to mirror surface in $X_{sa}Z_{sa}$-plane and X_{sa}-axis,
q_8 : angular offset of laser beam around Y_a-axis,
q_9 : angular offset of laser beam around X_a-axis,
q_{10} : x-offset of laser beam in S_a,
q_{11} : y-offset of laser beam in S_a,
q_{12} : offset angle of direct drive motor encoder (unit B)
q_{13} : linear offset parameter of mirror frame with respect to S_b, along X_b-axis,
q_{14} : angular offset of mirror frame S_{sb}, around X_b-axis,
q_{15} : offset angle of galvanomotor encoder (unit B),
q_{16} : distance between mirror surface and origin of frame S_b,
q_{17} : angle between projection of normal to mirror surface in X_{sb} Y_{sb} -plane and X_{sb} -axis,
q_{18} : angle between projection of normal to mirror surface in X_{sb} Z_{sb} -plane and X_{sb} -axis,
q_{19} : angular offset of laser beam around Y_b-axis,
q_{20} : angular offset of laser beam around X_b-axis,
q_{21} : x-offset of laser beam in S_b,
q_{22} : y-offset of laser beam in S_b,
q_{23} : roll parameter for transformation of S_b to S_h (which is chosen to coincide with S_a),
q_{24} : pitch parameter for transformation of S_b to S_h ,
q_{25} : yaw parameter for transformation of S_b to S_h ,
q_{26} : x-distance from S_b to S_h ,
q_{27} : y-distance from S_b to S_h ,
q_{28} : z-distance from S_b to S_h .

Due to the difference between the real system and the mathematical model, the computed point p does not coincide with the real target position. To remedy this, the model parameters are adapted until the two positions coincide, based upon the measurements of a series of exactly known target point positions (using a coordinate measuring machine). This procedure is called *calibration* of the measuring setup, and will be explained hereafter in more detail.

3.5.4. Calibration Criteria

Calibration of the measurement system is obtained through minimizing following error function:

$$\frac{1}{2}\sum_{j=1}^{m} f_j^2(\mathbf{q}, \boldsymbol{\theta}_j), \tag{44}$$

where

$q = (q_1, q_2, \ldots, q_{28})$ is the vector of model parameters to be determined,

$\theta = (\theta_1,...,\theta_j,...,\theta_m)T$ is the input to the model,
$m =$ number of measuring points,
$\theta_j = \theta_{1a}, \theta_{a2}, PSD_{xa}, PSD_{ya}, \theta_{1b}, \theta_{2b}, PSD_{xb}, PSD_{yb}),$

with

θ_{1a}, θ_{1b} the angular position of direct drive motors *A* and *B*,
θ_{2a}, θ_{2b} the angular position of galvanometers *A* and *B*,
PSD_{xa}, PSD_{xb}, the x-outputs of *PSDs A* and *B*,
PSD_{ya}, PSD_{yb}, the y-outputs of *PSDs A* and *B*,
f_j is the residue for measuring point *j*.

Residues f_j can be determined by three different methods:

Method of crossing beams:
In this method, the residue f_j is the minimum distance between the crossing beams, which is to be minimized. The advantage of this method is that measurement of m arbitrary points suffices to calibrate the system. The main drawback, however, which makes the method virtually useless, is that the requirement of intersecting beams does not guarantee that the beams really intersect.

Position method:
The model parameters can also be determined by requiring that the deviation between the target position obtained from the model and the real position (which is then the residue f_j) be minimal.

Combined method:
Both previous methods can be combined by requiring that the shortest distance between the outgoing laser beam from unit A and the real target point position be minimal and that the shortest distance between the computer laser beam form unit B and the position of the real target point be minimal. The residue f_j is then splitted in two residues:
- f_{jA}, the minimal distance between the position of the real target point and the outgoing laser beam from *A*,
- f_{jB}, the minimal distance between the position of the real target point and laser beam B.

This third method has been implemented in the final laser tracker. The method requires a reference grid of accurately known measurements points. A coordinate measuring machine has been used for this purpose.

3.5.5. Calibration Algorithm

Solving the calibration problem consists in searching a vector q of 28 model parameters q_i that minimizes the sum of squares of deviations f_j, accumulated over a set of m measuring points. Input to the model is a vector θ, defined under equation (44).

A lot of procedures are described in literature to solve such nonlinear minimization problems. For a goal function of the form:

$$\frac{1}{2}\sum_{j=1}^{m} f_j^2(\mathbf{q}, \theta_j), \tag{45}$$

the Levenberg-Marquardt algorithm provides a very elegant solution. It is a combination of a steepest descent and a Newton-Raphson algorithm. With the steepest descent method a minimum is searched. It is a robust method but converges slowly. Therefore the Newton-Raphson method approximates the function to be minimized in the neighborhood of the starting point by a quadratic function. The minimum of this quadratic is the new starting point.

In order to be able to apply the Levenberg-Marquardt algorithm, the original problem has to be linearized.

$$\text{minimize } ||F(q_{start} + F'(q_{start} P)|| \textit{ to } P$$

or

$$\text{minimize } ||\mathbf{F}(\mathbf{q}_{start} + \mathbf{J}.\mathbf{p})|| \tag{46}$$

This linearized least squares problem can now be solved by the Levenberg-Marquardt algorithm. For convergence it is important to start a good set of starting values q_{start}. This is realized by measuring directly as many parameters as possible. In [7] a detailed description is given of the different measuring methods used. For those parameters where no method was readily available, the default value zero was selected as starting value.

The large number of parameters in the model introduces some problems. Some parameters are insensitive for the reduction of the residues: they have to be changed considerably in order to have only a minor influence on the residue. These parameters are better removed from the model. Also those parameters, whose influence can be replaced by a linear combination of other parameters are to be removed from the model.

The Levenberg-Marquardt algorithm in fact solves a set of linear equations. The condition of this set is decisive for the condition of the model identification problem. A bad condition of the identification poblem is expressed by the singularity of the set of equations to be solved. The condition of the identification problem is determined by the condition of following set of linear equations:

$$J.p = - F, \tag{47}$$

where

J is the jacobian of the function to be minimized w.r.t. the model parameters,
p is the vector of proposed corrections of the model parameters,
F is the vector of deviations that are to be minimized.

In each iteration step of the Levenberg-Marquardt algorithm, the correction vector p has to be solved from equation (47). Condition problems arise when J is (nearly) singular. Physically, an ill-conditioned set of equations means that the measurements do not contain sufficient information to uniquely define the correction vector p. The singularity of J can have following effects:

- very large corrections on some parameters, which are from a physical standpoint very unlikely,
- identical or opposite corrections on parameters pairs.

Those effects can be reduced by following measures:

- improve the condition by measuring over a larger space,
- reduce the number of model parameters to those who are really independent.

The second method has been applied in the laser tracking system.

All the information required to select a set of independent and sensitive model parameters is contained in the jacobian matrix J. The technique of singular value decomposition (SVD) is applied to find this set of parameters.

The set of linear equations (47) is overdetermined and usually ill-conditioned, as explained before. The solution p is to be determined so that the residue $J\,p + F$ is minimal in the least squares sense. SDV is a suitable technique to solve least squares problems. By SVD, equation (47) can be written as:

$$(U.W.V^T)\,p = -F, \tag{48}$$

and the solution becomes:

$$p = -(U.W.V^T)^{-1}.F, \tag{49}$$

or

$$p = -V.diag(r^{-1}).U.F. \tag{50}$$

For all singular values r_j that are zero or that have a value in the vicinity of the machine resolution, $1/\sigma_j$ is put equal to zero. It can be proven that the resulting solution p minimizes the residue $Jp = F$. The selection of the threshold value below which the singular value can be put equal to zero is delicate and needs some special care [7].

The selection of a set of independent, sufficiently sensitive model parameters is thus done by means of a singular value decomposition of jacobian J. Suppose that W has r nonzero singular values and consequently $n - r$ zero values (or equivalent to zero). The rank of J is thus r. This means that J has r independent columns, and thus that the model has r independent model parameters.

From V^T, a set of independent column vectors can be determined that form a basis for the solution space. Because of the fact that each model parameter is coupled to a column vector of V^T, a set of independent model parameters can be derived form that set of independent column vectors. Therefore the following procedure can be followed:

- for each parameter i the norm $\|V_i^T.W\|$ is computed. The larger this norm, the more the parameter contributes to the reduction of the residue.
- all parameters are ordered in decreasing order of $\|V_i^T.W\|$. Those parameters with a contribution lower than a predefined minimum are omitted from the model.
- an independence test can be performed on the set of selected column vectors V_i^T, by requiring their mutual orthogonality [7].

Application of this method to the LASER-TREK system resulted in a reduction of the number of error parameters from 28 to 14 independent and sufficiently sensitive parameters. In this way, the problems with insensitive parameters, which cause trouble in the original model, are avoided, while the simple model results in the same accuracy.

3.5.6. Operational Data and System Accuracy

The performance of LASER-TREK has been extensively tested using a coordinate measuring machine (CMM). This CMM generates accurate reference position data. Tests have been executed in a measuring space of *300* x *300* x *250 mm*, laid-out according to Figure 24. The measurements and calibration results are summarized in table 3. From these date, some important conclusions can be drawn:

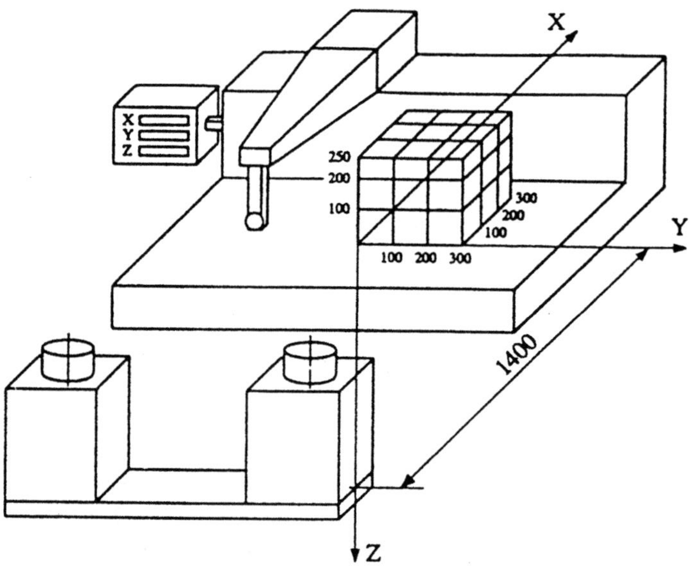

Figure 24: Calibration setup and calibration space

- The distance measurement accuracy is:
  . along X-axis: *0.25 - 0.28 %*,
  . along Y-axis: *0.06 - 0.14 %*,
  . along Z-axis: *0.03 - 0.06 %*.
- The resolution is:
  . along X-axis: +/- *0.04 mm*,
  . along Y - and Z-axis: +/- *0.01 mm*

As indicated in Figure 25 and 26, the measurement noise is:
- along X-axis: +/- *0.04 mm*,

- along Y- and Z-axis: *+/- 0.01 mm.*

The discrepancy between the accuracies along the *X*-axis with respect to those along *Y* - and *Z*-axis are completely due to the condition of the measurement steup. Due to the limited opening angle of the corner cube retro-reflector, the two deflection units have to be positioned relatively close to each other. The use of a cat's eye would enhance the measurement conditions considerably.

It has to be said that above remarkable results are *absolute* measurements. It is clear that the accuracy is much higher when comparative measurements are performed.

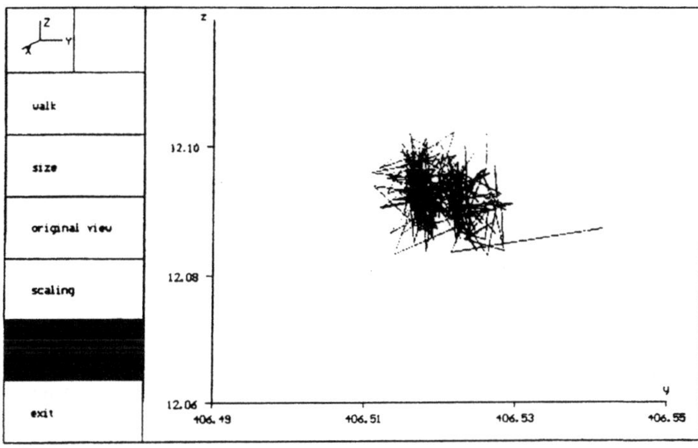

Figure 25: Measurement noise in YZ-plane

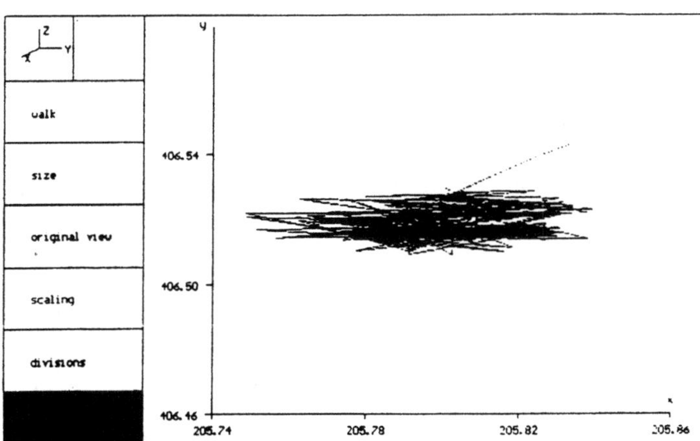

Figure 26: Measurement noise in XY-plane

y=0, z=0

X-axis (mm)	d_m (mm)	Δ_t (mm)	Δ_{st} (mm)	Error (%)
0	0	0	0	-
100	99.712	-0.288	-0.288	-0.28
200	199.432	-0.567	-0.279	-0.28
300	299.235	-0.764	-0.179	-0.25

x=1700, z=0

Y-axis (mm)	d_m (mm)	Δ_t (mm)	Δ_{st} (mm)	Error (%)
0	0	0	0	-
100	99.860	-0.140	-0.140	-0.14
200	199.716	-0.284	-0.144	-0.14
300	299.605	-0.395	-0.111	-0.13

x=1400, z=0

Y-axis (mm)	d_m (mm)	Δ_t (mm)	Δ_{st} (mm)	Error (%)
0	0	0	0	-
100	99.895	-0.104	-0.104	-0.10
200	199.848	-0.152	-0.048	-0.07
300	299.793	-0.207	-0.055	-0.06

x=1700, y=0

Z-axis (mm)	d_m (mm)	Δ_t (mm)	Δ_{st} (mm)	Error (%)
0	0	0	0	-
100	99.934	-0.066	-0.066	-0.06
200	200.007	0.059	0.007	0.03
250	250.100	0.100	0.041	0.033

x=1400, y=0

Z-axis (mm)	d_m (mm)	Δ_t (mm)	Δ_{st} (mm)	Error (%)
0	0	0	0	-
100	99.943	-0.057	-0.057	-0.05
200	200.014	0.014	0.070	0.005
250	250.085	0.085	0.071	0.035

d_m = distance measured
Δ_t = total measuring error
Δ_{st} = error for step of 100 mm

Table 3: Measuring results with calibrated LASER-TREK system

Bibliography

[1] Fresonke D.A., Hernandez, E., "Deflection Prediction for Serial Manipulators", Proceedings of the 1988 International Conference on Robotics and Automation, Philadelphia, pp. 482-487, 1988.

[2] Craig J. J., "Introduction to Robotics: Mechanics and Control", Addison-Wesley, Reading, MA, 1986.

[3] Van Brussel et al., "Identification and Control of Flexible Structures", Proceedings of the International Conference on Advanced Mechatronics, JSPE, Tokyo, pp.347-352, 1989.

[4] ANSI/RIA R15.05, "Evaluation of the Static Performance Characteristics for the Comparison of Robots and Robotic Systems", Revision 10.

[5] ISO 9283, "Manipulating Industrial Robots. Performance Criteria and Related Testing Methods".

[6] VDI 2861, Blatt 2, Entwurf: "Kenngrossen fur Industrieroboter. Einsatz spezifischer Kenngroessen", 1987.

[7] Van Waeyenberge B., Wijnants J., "Optimalisation of the Calibration of a 3D Laser Tracking System", Thesis Division PMA, K.U.Leuven, 1991 (in Dutch).

PART III

Industrial Application of the SACODY Robot Control

Chapter 7

Controller Design for a High-Performance Servo Level

H. Wunderlich, H. Walter
AEG AG Boeblingen, Germany

1. Introduction

1.1. Requirements for a Robot Control with a High-Performance Servo Level

Today many industrial robots with various mechanical structures and different control configurations are installed in flexible manufacturing systems mostly in the automotive industry. Although they are introduced nearly since 2 - 3 decades, there is still a strong demand for higher accuracy, active damping of flexible mechanical structures and an increase in flexibility of the whole system. Additional demands for robot systems in the field of factory automation are to reduce costs for example with a shorter cycletime of the production process and to integrate robots in new and complex technological applications like mounting, cutting etc. To solve these problems a robot control must be extended to integrate new control procedures and sensor signal processing. The design of hardware and software modules must lead to a higher flexibility of a robot control. Such a flexible robot control can be used in different configurations for various robot systems, drive systems and applications [1].

The focus of SACODY project is the servo level of a robot control. Corresponding to this task Table 1 gives an overview of the main requirements for such a high-performance servo level.

- high position accuracy
- short positioning time
- active damping for a flexible mechanical structure
- high computation power for advanced control algorithms
- flexibility in respect of different servo drive systems
- flexibility in respect of different encoder interfaces
- extension in respect of different sensor signals
- integration into the overall robot control design

Table 1: Requirements for a High-performance Servo Level

1.2. Tasks of AEG in SACODY Project

The main task of AEG during SACODY project was to develop a high-performance servo level for position control and sensor signal processing.

To fulfil the requirements of SACODY project, the development of the new position and sensor control units LAR 830/286 and LAR 831/186 with additional interface boards was realized in hard-and software. This new servo level environment was integrated into the overall design of the AEG robot control.

In a next step, the robot control was installed at a KUKA robot. As a result of the theoretical and practical research of this project, advanced algorithms for position control as described in chapters 2 and 8 were incorporated into the controller. For measurements and tests, the robot system and the controller were integrated in an industrial cell for spot welding (see chapter 9). A summary of the several tasks for this project is given in Table 2.

- General task
  New servo level environment in order to incorporate advanced algorithms for
  position control and
  sensor signal processing
- Hardware development
  * position control unit LAR 830/286
  * programmable interface- and sensor-unit LAR 831/186 with additional analog input signals
  * encoder interface for 9 axes
  * initialization and synchronization
  * connections to different bus-systems (dual port memory) and interfaces (plugs)
- Software development
  * coordinate transformation for KUKA robot with central angle hand
  * adaption of AEG position control algorithms
  * extension of sensor signal processing for analog I/O, digital I/O and encoder interfaces
  * task organization in respect of bus systems
  * organization and adjustment of machine parameters
- Installation and measurements

Table 2: Development of AEG Controller

2. System Architecture of AEG Robot Control

2.1. System Software and Control Functions

To ensure a better understanding of the development and integration of the new servo level which have been realized during this project an overview of the system architecture of the AEG robot control is given in this preliminary chapter.

The performance of a robot control is mainly given by the level of control functions.

They are realized in the software of the system and servo level of the global system control, see Figure 1.

Via the system manager, there are interfaces to the operating level (teach panel) and the CIM level (cell controller). There are other interfaces to a robot. These can be connections to a technology specific tool and the drive system. The interfaces to the sensor system can be separated in the internal measurement systems at the drives and additional external sensors.

The standard control functions of the system level also describe the possibilities of the customer how to use a robot system (Figure 1) [2]:

- operation functions,
- motion planning and interpolation,
- logic unit or internal PLC,
- global sensor processing,
- diagnostic functions
- handling of machine parameters and
- data-, program- and interface-management.

From the system level the demand values are transfered to the servo level, which includes the following procedures:

- local axis control,
- position control, active damping and
- sensor signal processing.

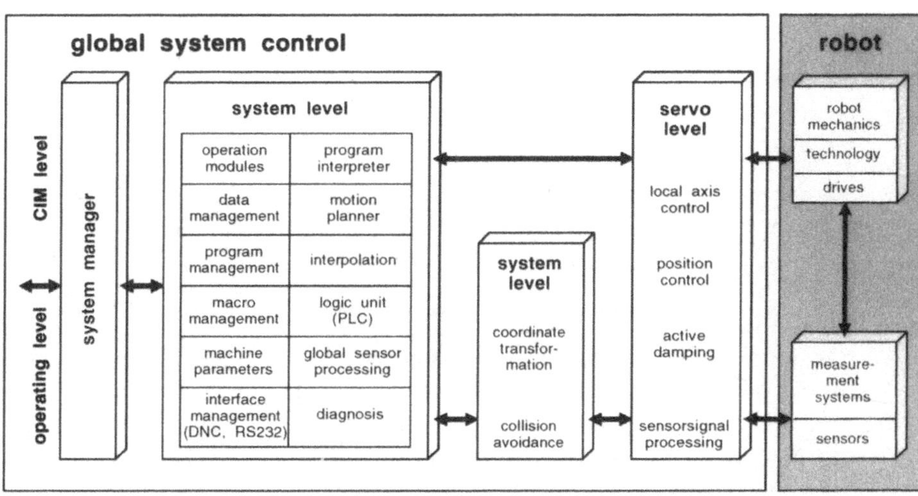

Figure 1: System Software of AEG Robot Control

By the help of the clear structure of the menu technique, the operator is guided to handle the various control functions and operation modes [3]:

- commissioning of machine parameters,
- visualization of data and messages,
- setting up main- and subprograms,
- setting up PLC-programs,
- teaching of points and trajectories,
- testing programs,
- running programs and
- archiving.

The main control functions of the AEG robot control are listed in Table 3. The programming language DOROB of the AEG robot control is a userfriendly tool to define the demanded task in a number of programs. It is possible to set up up to 100 main- and subprograms. A program may include all informations for a trajectory and logic functions. In addition to a teach panel an offline programming system based on a comfortable window-technique can be used for programming [2,3].

The coordinate transformation algorithms enable the operator to manipulate a machine with up to 12 axes and a complex kinematic structure in JOINT-, WORLD-, TOOL- and FRAME-coordinates. In addition to a simple point-to-point movement the powerful interpolation procedure allows to realize a linear, a parabolic or a circular

Control Functions

- ■ powerful program language DOROB
- ■ structured programming with main- and subprograms
- ■ offline programming
- ■ systems of coordinates
  - * JOINT, WORLD, TOOL, FRAME
- ■ path planning and interpolation
  - * PTP (Point-To-Point)
  - * CP (Continuous Path)
    - linear
    - defined slope (parabola 2nd order)
    - parabolic (parabola 3rd order)
    - circular
- ■ system- and program-PLC
  - * 2000 instructions
  - * 32 registers and 32 counters
  - * 540 markers
- ■ synchronous and asynchronous movement of up to 12 axes
- ■ override-function for trajectories via a 3 dimensional joystick
- ■ macros
  - * collision avaidance
  - * deflection compensation
  - * technology functions

Table 3: Technical Data of AEG Robot Control Software

path. Important for the programming is the internal PLC, which allows all necessary logical and arithmetical operations. The calculation of the motion planner assures that the several PLC- and sensor-functions are exactly activated at the demanded position. With the help of the 3-dimensional joystick or an external sensor, it is possible to realize an online override function for programmed trajectories.

As a so-called technology carrier a robot system needs several specific functions for example for welding, mounting or painting. Such functions are implemented in special macros in the AEG robot control. These macros are integrated in the overall system software. They can be handled by the user as a robot program. The parameters for such a macro can be set up with the operating device.

2.2. Hardware Configuration

The AEG robot control is a modular multiprocessor controller for industrial robots. It may include up to five microprocessors and is mounted in a 19" rack. The hardware

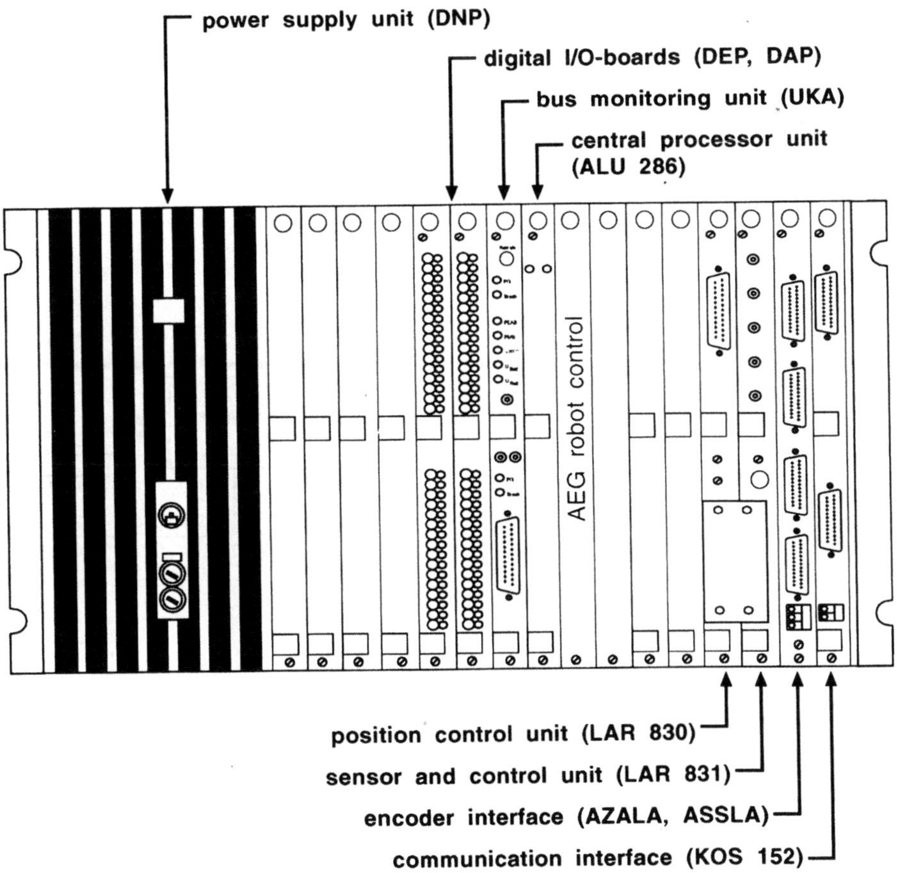

Figure 2: Overall Design of AEG Robot Control Hardware

is configurated with standard components based on the automation system MODICON of AEG. Depending on various machines or applications the standard robot control can be extended or adapted either in hardware components or by means of software modules. The configuration used in SACODY project is shown in Figure 2.

The internal structure of the control system is described in Figure 3. As a master the main processor board CPU/ALU is connected to the internal parallel memory bus (PMB).

Additionally the ALU is directly connected to the PEAB, the parallel I/0 bus. At the PEAB the boards for the robot control inputs and outputs are connected to handle the I/0 signals for the internal PLC of the robot control.

The boards for communication and position control are coupled to the PMB. The communication interface board KOS is equipped with several serial RS 232 interfaces to handle the teach panel and to communicate with a PC for archiving, DNC-interfacing and offline-programming.

For position control and sensor signal processing there are several possibilities to use different hardware components depending on the control task and the needed computation power. In SACODY project the solution with the LAR 830/286 and LAR 831/186 for a high-performance servo level is integrated, see Figure 3. This structure was chosen to fulfil the high demands for sensor signal processing and position control with modern an powerful control algorithms.

For alternative applications and in a version with reduced costs the central control unit LAR 830/286 may be taken off and the control unit LAR 831/186 can be connected directly to the PMB system bus, see Figure 4. This configuration provides

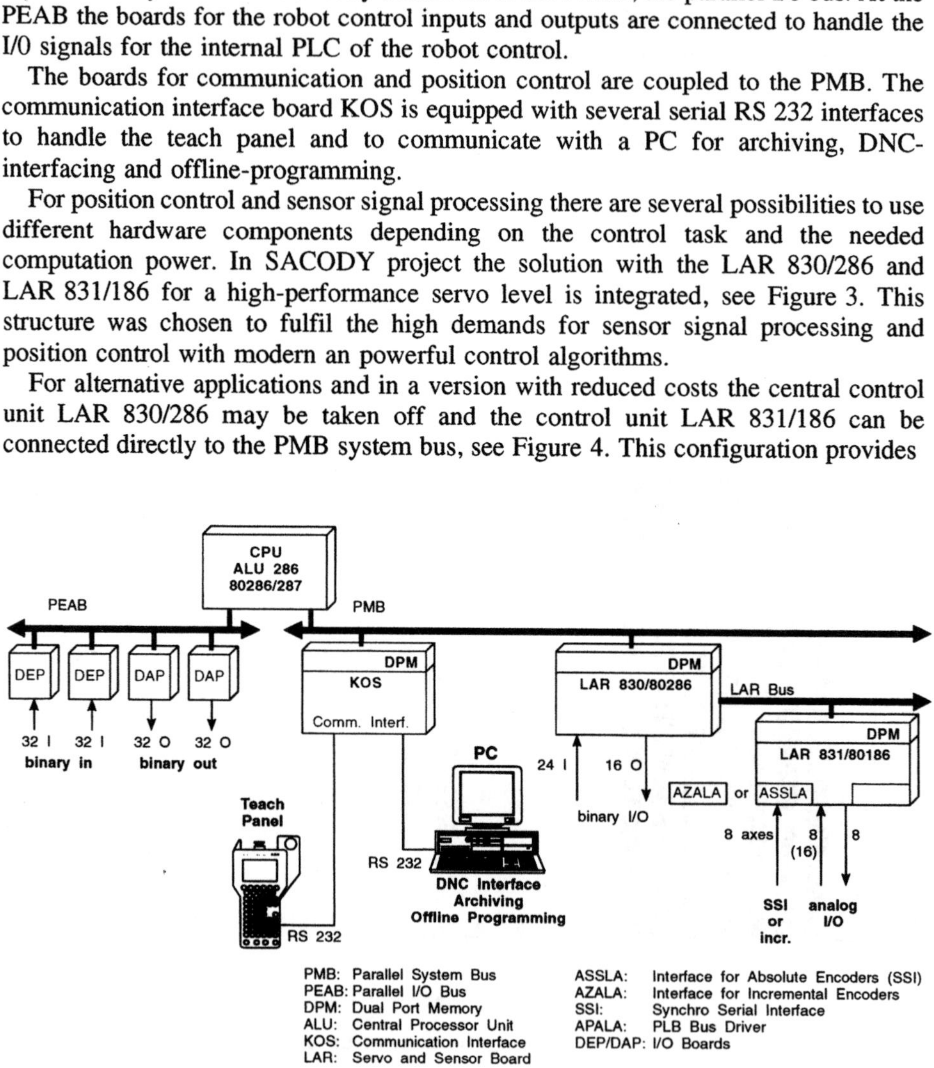

Figure 3: AEG Robot Control with High-Performance Servo Level

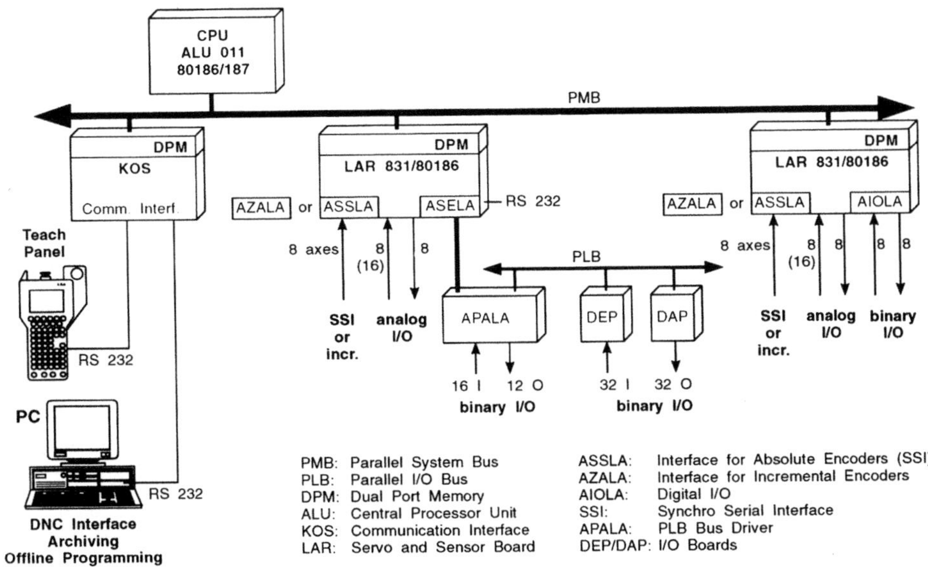

Figure 4: Structure of AEG Robot Control IRC 500 for up to 12 Axes

Hardware

- modular multi processor system
- DNC-interface
- teach panel with 3 dim. joystick
- connections and plugs at the front
- position control and sensor boards
  - * high computation power
  - * analog input: 12 bit resolution
    8-16 double ended, +/- 10V
    16-32 single ended, 0-10V
  - * analog output: 13 bit resolution
    8-16, +/- 10V
  - * digital input
    16-40, 24V
  - * digital output
    12-30, 24V
- encoder interface boards
  - * incremental encoders
  - * absolute encoders (SSI-interface)
  - * double resolver system
- digital I/O-boards for PLC
  - * 4 slots each for 32 dig. I/O, 24V

Table 4: Technical Data of AEG Robot Control Hardware

less computation power for control algorithms, but with two control units LAR 831/186 at the PMB it is possible to set up a controller for a robot system with up to 12 servo axes.

In the configuration shown in Figure 4 the I/0 bus PEAB is replaced by the PLB. The connection to the PLB is realized by the APALA board. This solution reduces costs and ensures a fast processing of I/0 signals in the cycle time of the servo level. Figure 5 is a photo of the AEG robot control for 12 axes referring to the internal structure given in Figure 4.

A summary of the technical data and hardware components is given in Table 4 [4].

Figure 5: AEG Robot Control IRC 500

3. Development of the Controller

3.1. Hardware for Servo Level

The first steps in the development of a new control unit are the specification and the design phase. Refering to the focus of SACODY project the following items were taken into account:

- requirements of the R & D work plan for advanced control algorithms,
- requirements for industrial FMS robot systems,
- modular design and flexibility in software and hardware configurations [1],
- integration into the overall control design and
- costs of the controller.

As a result of this first step the following boards for the high-performance servo level were specified:

- position control unit LAR 830/286,
- sensor and control unit LAR 831/186 and
- interface boards.

With the realization of this boards the robot control provides a high computation power for control algorithms and sensorsignal processing. All the calculations at the servo level can be carried out in a short cycle time of about *2 - 4 ms*. The servo level is extended to handle up to 12 servo axes as well as additional digital and analog input and output signals. The in-and outputs can be configurated depending on the task of a robot. The specified boards are now described in more details.

3.1.1. Position Control Unit LAR 830/286

The LAR 830/286 is a powerful computer board with an Intel 80286 processor and additional digital I/0. It is connected to the system bus PMB (see Figures 3 and 4). Another interface is the LAR bus for the connection to one board LAR 831/186. Algorithms for central position- and sensor signal-control may be implemented on the board LAR 830/286.

The technical data are listed in Table 5 and a photo of the final board is shown in Figure 6.

3.1.2. Sensor and Control Unit LAR 831/186

The LAR 831/186 is an intelligent analog I/0 board with the Intel processor 80C186. It is equipped with 8 analog outputs for demand values. As an option the board can be extended by 8/16 analog inputs with a high resolution. The board can be connected either to the LAR 830/286 via the LAR bus or directly to the PMB (see Figures 3 and 4). Additionally there is a plug for the connection with several interface boards. The technical data are listed in Table 6.

Technical Data	
Processor	: Intel 80286, 10 MHz
EPROM	: 256 KB
RAM	: 256 KB
Dual Port Memory	: 4 KB to connect PMB Bus
LAR Bus	: coupling to control unit LAR 831/186
Binary Inputs	
number	: 24
voltage	: 24 VDC
input current	: 4 mA
Binary Outputs	
number	: 18
voltage	: 24 VDC short circuit proof
nominal current	: 100 mA

Table 5: Technical Data of Position Control Unit LAR 830/286

Figure 6: Position Control Unit LAR 830/286

Technical Data	
Processor	: Intel 80C186, 12 MHz
EPROM	: 64/128 KB
RAM	: 16/64 KB
Dual Port Memory	: 4 KB to connect either PMB or LAR bus
Analog Inputs	
number of channels	: 16 single-ended or 8 differential inputs
resolution	: 12 bit
range	: unipolar, 0...10V bipolar, +/-5V, +/-10V
amplifier gain	: 1,2,10 adjustable
filter	: analog active low pass filter 2nd order (i. e. Bessel-filter)
Analog Outputs	
number of channels	: 8
resolution	: 12 bit + sign
output voltage	: +/-10V
output current	: 10mA
Additional Interface Boards	

Table 6: Technical Data of Sensor and Control Unit LAR 831/186

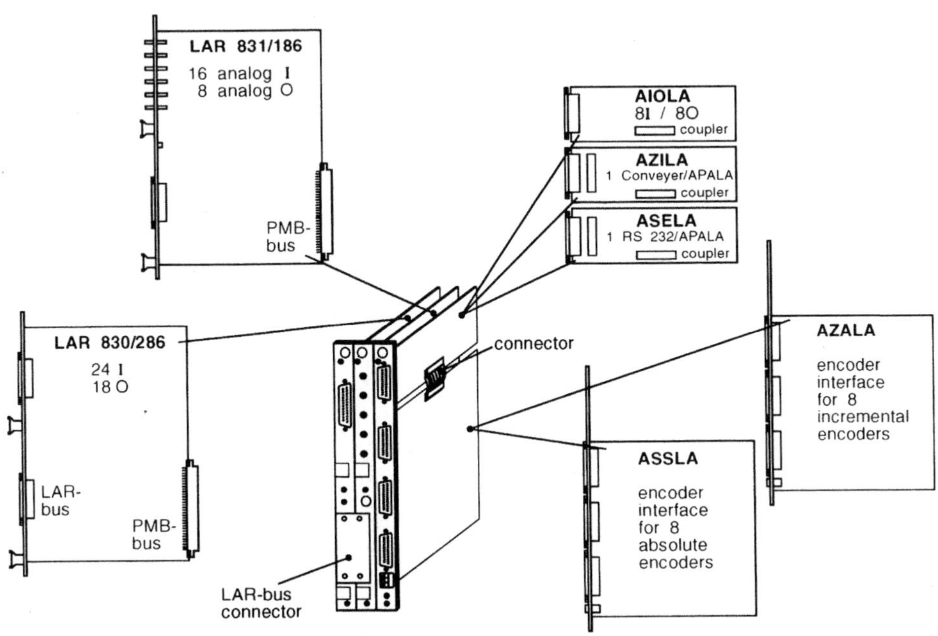

Figure 7: Hardware Configurations for Servo Level

3.1.3. Interface Boards for LAR 831/186

The following interfaces for different encoders, sensors and several other functions which can be connected to the LAR 831/186 are realized. One encoder interface can be used for maximal 9 encoders.

- interface for incremental encoders (AZALA)
- interface for absolute encoders with SSI protocol (ASSLA)
- board for additional digital I/0 of drives (AIOLA)
- board for one additional counter or conveyor measurement system (AZILA)
- RS 232 serial interface and connection to the board APALA (ASELA)
- interface to PLB I/O bus (APALA).

3.1.4. Hardware for Servo Level

With the previous described components different configurations as shown in Figure 7 can be set up for the servo level of the robot control. Special for the high-performance servo level in SACODY project the configuration in Figure 8 is used. It consists of the LAR 830/286, the LAR 831/186, the incremental encoder interface AZALA and the board ASELA (see Figure 7).

Figure 8: Position and Sensor Controller

3.2. Software for Servo Level

In addition to the development of new hardware boards it was necessary to set up a new software for the servo level related to the specifications in SACODY project. As a part of the workplan the conventional software was extended and implemented in the new boards. By other partners the advanced control algorithms were adapted and installed in the same boards.

Figure 9 shows the complete robot system with the system level and the servo level of the conventional controller. The robot control is connected to the electrical, mechanical- and measurement-systems of the robot.

In SACODY project, there is an important difference to standard robots: At the three main axes additional encoders are mounted behind the gears of the drive systems. This measurement is important to realize and test control algorithms with active damping of flexible mechanical structures.

From the system level the position demand for each axis is given to the servo level. The communication between system and servo level is realized in the cycle time of the system level, which is approximately *30 ms* for a six axes robot. The response from the servo- to the system-level are actual position values as well as control and monitoring informations.

The feedback control at the servo level is a conventional procedure with a cascade structure of digital position control and ananalog speed and current control. The digital position control with a feedforward control of speed is realized on the LAR 830/286 for up to 8 axes and the analog control is a task of the servo amplifier.

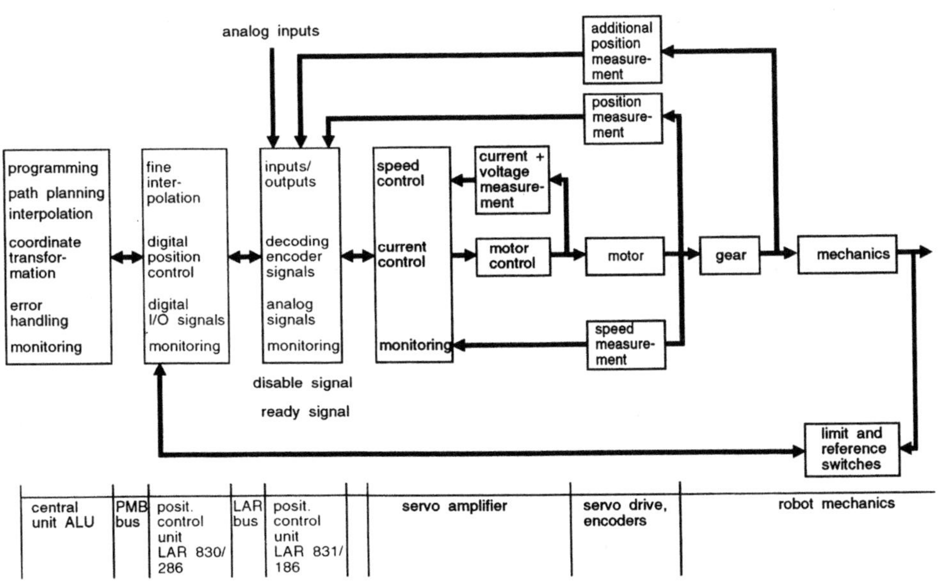

Figure 9: Robot System

To achieve a smooth and accurate positioning, the cycle time of the servo level is *2 - 4 ms*. Therefore a digital filtering of the position demand values from the system level is necessary. This is realized in a special fine interpolation on the servo level.

The standard servo level software consists of different modular software modules which are running sequencely as it is shown in Figure 10.

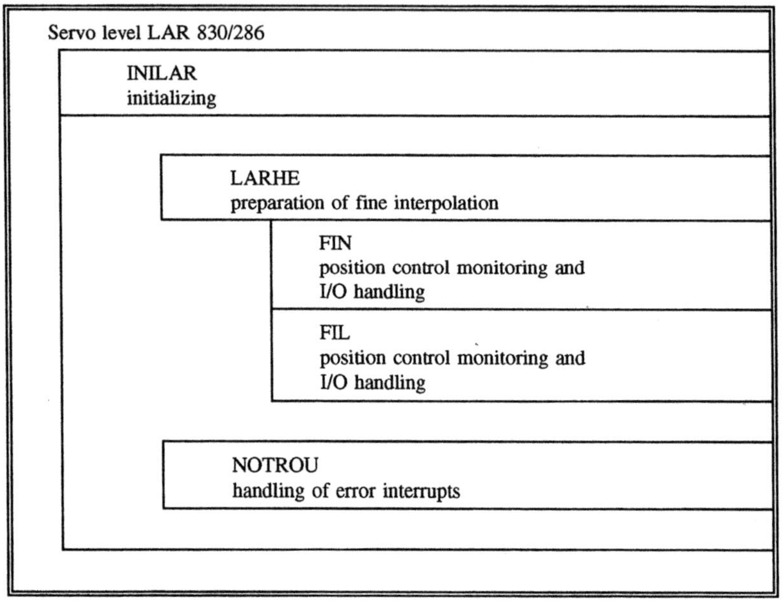

Figure 10: Software of LAR 830/286

To enable the calculations on the unit LAR 830/286 an interrupt controlled process management is integrated. The software modul LARHE for the preparation of the fine interpolation is running to prepare the values for the next system cycle and is interrupted every *2.5 ms* by a hardware interrupt from the board LAR 831/186. With this interrupt LARHE is stopped and the moduls FIN and FIL are started. After finishing their task the module LARHE continues running. In Figure 11 the blockdiagram of LAR 830/286 shows the internal handling and flow of data and control signals which are used for position control and the interfaces to PMB and LAR bus.

The software was implemented and sucessfully tested with the KUKA robot. The kernel of this servo level software was used to implement the advanced control algorithms of SACODY project. This part was realized by other partners and is described in the following chapters.

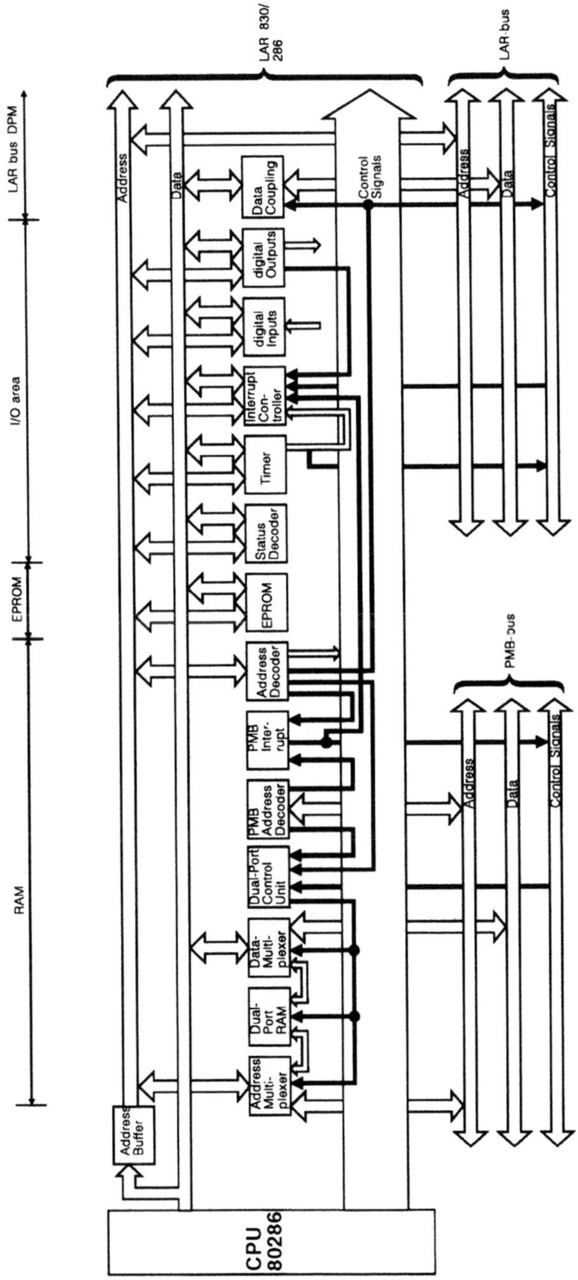

Figure 11: Block Diagramm of LAR 830/286

4. Summary and Outlook

A new servo level in hardware and software has been developed during SACODY project. The successful tests and measurements demonstrate, that there is an evident improvement in position control and sensor signal processing. Using the results for industrial robot systems there will be an increase in:

- PRODUCTIVITY
  . high quality by high accuracy
  . short cycle time of the production process
  . less costs for the robot mechanic
- NEW ROBOT SYSTEMS
  . flexible mechanical structure
  . robot kinematics with more than 6 axes
  . flexibility in adaption of different servo drive systems
- NEW APPLICATIONS
  . high accuracy in positioning and high velocity
  . integration of sensors
  . integration of algorithms for technology functions.

Figure 12: AEG Robot Control for Large Manipulators

The focus of SACODY project was an online dynamic compensation for a high-performance FMS robot. With the positive results of this project it is possible to realize an important spin off: The application of the AEG robot control with the new high-performance servo level to large manipulators with a redundant kinematic, a flexible mechanical structure and hydraulic servo drives, see Figure 12 [1,5,6].

The positive results of SACODY project are encouraging to continue research and development for a robot control in the crucial points mentioned in Figure 13.

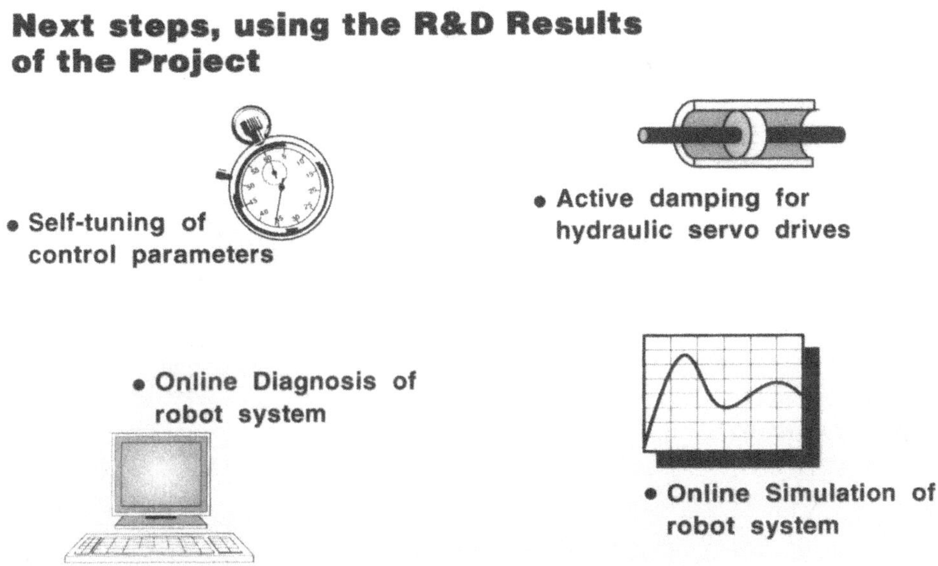

Figure 13: Outlook

Bibliography

[1] Wunderlich, H.: Flexible Robot Control for the Building Industry. 8th International Symposium on Automation and Robotics in Construction (I.S.A.R.C), Stuttgart, June 1991.

[2] AEG: AEG robot control. Bedien- und Programmier-Handbuch, 1991.

[3] Wunderlich, H.; Walter, H.: Hard- und Software der zentralen Bedienschnittstelle. Fachberichte Messen, Steuern, Regeln 24. Fortschrittliche Robotersteuerungstechnik. Hrsg. Schmid, D., Springer-Verlag, Berlin, Heidelberg, 1991, pp. 220-226.

[4] AEG: AEG robot control. Inbetriebnahme- und Hardware-Handbuch, 1991.

[5] Schraft, R.D.; Bodenmüller, H.; Benckert, H.: Entwicklung eines mobilen Grobroboters. F+H Fördern und Heben 39 (1989), Nr. 4, pp. 333-336.

[6] AEG AG: Flexibilität in der Handhabungstechnik, Technik Magazin 2/90, pp. 19-21.

Chapter 8

Design and Implementation of an Antivibration Robot Control Software

Joël Deniard, Jean-Luc Faillot
Bertin et Cie

Jan Swevers, Dirk Torfs
K.U.Leuven

1. Introduction

The different tasks carried out during the course of the SACODY project, and devised in the previous chapters, have contributed to the definition of a complete methodology for the design of high performance control for flexible mechanisms.

Although the problem of controlling flexible manipulators was at the beginning of SACODY an unsolved problem essentially dealt with at an academic level, the project was orientated towards a final demonstrator involving only industrial hardware in order to provide realistic results exploitable in the industry within a short leadtime. The hardware of this demonstrator consisted of the robot KUKA IR160 and of the AEG robot controller prototype described in the previous chapters.

This chapter focuses on the development and the integration in this hardware of the antivibration servo control software, that have been carried out in two steps in collaboration between the teams of KATHOLIEKE UNIVERSITEIT LEUVEN and BERTIN:

- First, the control has been designed by the K.U. LEUVEN and implemented on a laboratory multiprocessor, which provided all the flexibility necessary to test the functionalities of the new control algorithms designed after information on the robot modal behaviour released by the identification and the dynamic modelling phases. This design phase is reported section 2.
- Then, the resulting algorithms have been adapted by BERTIN, in order to cope with the implementation constraints specific to the AEG industrial prototype. Simulations of the controlled robot played again a important role for performing all the trade-offs necessary for the integration of the software in the target controller. This final integration task is reported section 3.

2. Design and Preliminary Implementation

2.1. Scope of the Developments

This section describes the design and the test of the antivibration control algorithms described in chapter 2 on an industrial robot KUKA IR 160/60. The algorithms developed present two level of modifications of a standard robot controller for:

- improving the trajectory generation,
- controlling the flexibilities of the first three joints.

The more advanced flexibility controller consists of a linear position-dependent state feedback controller for the first axis, and a nonlinear decoupling controller for the second and third axis (see chapter 2).

The modifications brought to the conventional controller structure result in a large improvement of the dynamic robot performance. A comparison of the performance of two improved robot controllers with the standard robot controller during tests based on the test-standards of ISO (Industrial Organisation for Standardization), and RIA (Robotic Industries Association), shows the improved dynamic performance.

These tests also show that the improved trajectory generation gives the largest contribution to the improvement of the performance, but also that at very high velocities and accelerations, there is a significant difference between a flexible controller and a classical PID controller. The preliminary antivibration controller exhibits also a better performance with respect to tracking accuracy and steady-state accuracy due to compensation of gravity.

This section decsribes the modelling, the identification, the design of the linear position-dependent state-feedback controller and the nonlinear decoupling controller, and overviews the experimental results obtained with this intermediate controller structure.

2.2. Description of the Test Setup

Figure 1 shows the set up for one of the tests. The robot is a KUKA IR 160/60 industrial robot with an *400 mm* arm extension: it has 6 axes, a *3.15 m* range, and is equipped with a spot welding tool of *45 kg*. Three extra encoders are attached to the first three links of the robot. They measure the link position besides the motor position, which is measured with an encoder mounted on the motor shaft. The difference between the motor encoder signal and the corresponding extra encoder signal corresponds to the deformation of the joint. An experimental modal analysis of this robot has shown that the main flexibility in this robot is caused by the first three joints [1], such that these extra encoders suffice to measure the flexibility. A digitizing pen and tablet together with the software package RODYM, which has been developed in this project (see chapter 6) [2], measure the end effector movements and positioning. The digitizing pen is attached to the tool of the robot.

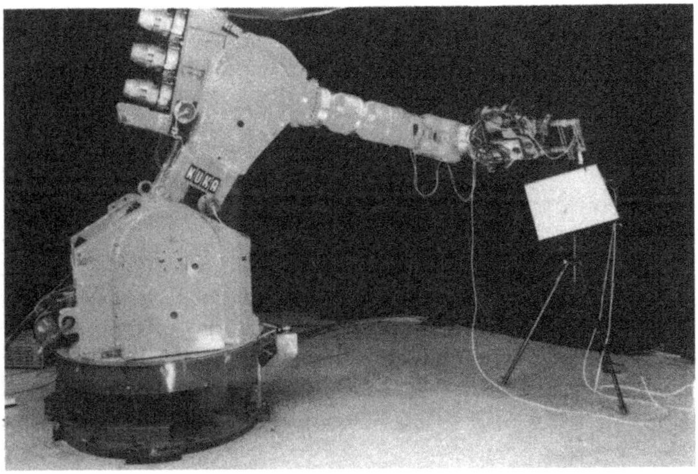

Figure 1: The KUKA IR160/60 equipped with its spot welding gun

The improved trajectory generation and the control algorithms are programmed on a VME-system. This system can read encoder signals and send out analog voltages in the range of *+/- 10 volt*. The power supply of the motors linearly converts the voltages into motor currents, when the power supply is working in torque command, or motor velocities, when it is working in velocity command. The command mode can be altered for each axis separately.

2.3. Linear Position-Dependent State-Feedback Control of the First Axis

2.3.1. Modelling and Identification

The controller for the first axis is based on 4 different fourth-order discrete-time linear models of the robot. Each of these models has been identified experimentally [3] for a certain configuration of axes 2 and 3. These configurations lie between the fully extended configuration and fully contracted configuration of the robot. The models take into account the flexibility of the first joint with one resonance and one anti-resonance frequency.

The identification of each fourth-order state-space models is based on the identification of two discrete-time transfer functions, as explained in chapter 2.

2.3.2. Control Design

The control signal $u[k]$ for the first axis is a weighted sum of four control signals $u_j[k]$.

$$u[k] = \Sigma w_j[k] u_j[k] \qquad (1)$$

$w_j[k]$ is the weight for $u_j[k]$. The control signals $u_j[k]$ consist of a feedforward and a feedback of the difference between the desired and the measured states:

$$u_j[k] = u_{ffj}[k] - K_{fbj}\,(x[k] - x_{dj}[k]) \tag{2}$$

The design of the state-feedback controllers and the calculation of the feedforward signals are based on discrete-time state-space models. The construction of the state-space models from the identified transfer functions and the design of the feedback and feedforward controllers is explained in chapter 2.

Each of these state-feedback controllers is stable in the whole working space of the robot, and gives a good dynamic behaviour in an area around the robot configuration for which they are designed. The weights $w_j[k]$ are functions of the distance of the end effector from the first axis of rotation. Figure 2 shows the four weights $w_j[k]$ as a function of $d[k]$. In this figure, d_1, d_2, d_3 and d_4 denote the distances of the end effector from the first axis of rotation, corresponding to the four robot configurations in which the robot is identified.

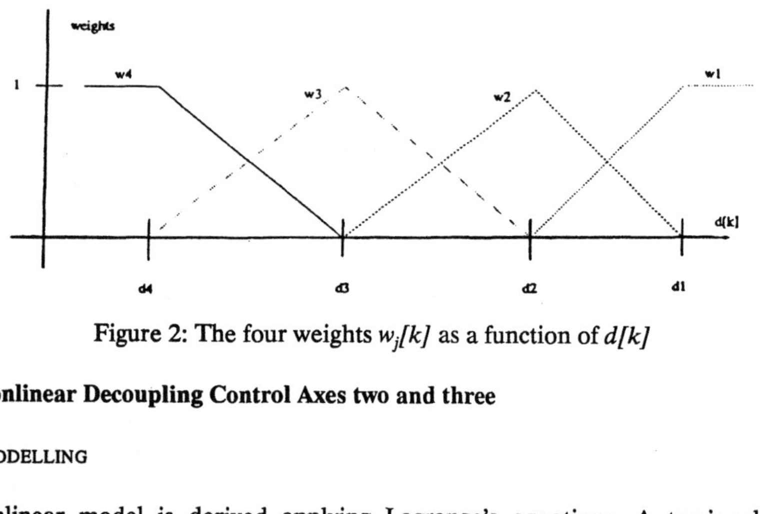

Figure 2: The four weights $w_j[k]$ as a function of $d[k]$

2.4. Nonlinear Decoupling Control Axes two and three

2.4.1. Modelling

The nonlinear model is derived applying Lagrange's equations. A torsional spring models the elastic coupling between the joints and the links.

$$(A_{22} + 2A_c cos\theta_3)\ddot{\theta}_2 + (I_{z3} + A_c cos\theta_3)\ddot{\theta}_3 + B_2(\theta_2,\theta_3) + K_2(\theta_2 - \frac{\Phi_2}{r_2}) = 0 \tag{3}$$

$$(I_{z3} + A_c cos\theta_3)\ddot{\theta}_2 + I_{z3}\ddot{\theta}_3 + B_3(\theta_2,\theta_3) + K_3(\theta_3 - \frac{\Phi_3}{r_3}) = 0 \tag{4}$$

$$I_{m2}\ddot{\Phi}_2 + T_{f2} - \frac{K_2}{r_2}(\theta_2 - \frac{\Phi_2}{r_2}) = T_2 \tag{5}$$

$$I_{m3}\ddot{\Phi}_3 + T_{f3} - \frac{K_3}{r_3}(\theta_3 - \frac{\Phi_3}{r_3}) = T_3 \tag{6}$$

where:

$$B_2(\theta_2,\theta_3) = -m_3 g x_{3c} \cos(\theta_2 + \theta_3) - (m_3 g l_2 + m_2 g x_{2c})\cos(\theta_2) - A_c \dot{\theta}_3 (2\dot{\theta}_2 + \dot{\theta}_3) \sin(\theta_3) - T_{comp}$$

$$B_3(\theta_2,\theta_3) = -m_3 g x_{3c} \cos(\theta_2 + \theta_3) + A_c \dot{\theta}_2^2 \sin(\theta_3)$$

and:

θ_2, θ_3 are the angles of rotation of the second and third actuators,
Φ_2, Φ_3 are the angles of rotation of the second and third links,
T_2, T_3 are the torques of motors 2 and 3,
r_2, r_3 are the gear ratios of transmissions 2 and 3,
I_{m2}, I_{m3} are the inertia of rotors 2 and 3,
I_{z3} is the moment of inertia of the third link,
T_{f2}, T_{f3} are the friction torques of motors 2 and 3,
K_2, K_3 are the spring constants of joints 2 and 3,
$A_{22} = I_{z2} + I_{z3} + m_3 l_2^2$
$A_c = m_3 l_2 x_{3c}$
T_{comp} is the torque compensating the gravity,
m_i is the mass of the link i,
x_i is the position of the mass center of the link i.

The second link position is measured with respect to the horizontal axis. The third link position is measured with respect to the second link. The motor friction consists of linear viscous friction and Coulomb friction.

2.4.2. Identification

The identification of the parameters in this model is performed in differnet steps: First, the parameters related to the rigid body model are estimated. Then the spring constants are estimated by using static, quasi-static, noise measurements and stepped sine measurements. The identification procedure is explained chapter 2.

Identification of inertial and gravitational coefficients
The identification method requires motion of a single axis, while the positions of the others are fixed. A test motion with high accelerations but small changes in position is chosen. In this way there are no coupling torques acting on the moving axis, and the gravitational forces can be regarded as constant. The main component of the motor torque can be related to the effective inertia of the moving axis and to the constant gravitational forces. The same measurement enables to determine the coupling inertias, by considering the control torques needed to maintain the fixed position at the other axes. Repeating the procedure in different positions, sets of unknown inertial and gravitational coefficients are obtained as a function of the joint position. The solution of this set of equations gives A_{22}, A_c and I_{z3}.

Identification of the spring constants

- Static measurements (in four different positions): the end effector (without spot welding tool) is loaded with a mass $m_l = 91.5\,kg$. The deformation before and after loading is measured. From the difference between these two deformations Δq_2 and Δq_3, the spring constants are calculated:

$$K_2 = [\, m\, g\, l_3 \cos(\theta_2 + \theta_3) + m\, g\, l_2 \cos(\theta_2)\,] \,/\, \Delta q_2$$
$$K_3 = m\, g\, l_3 \cos(\theta_2 + \theta_3) \,/\, \Delta q_3$$

The same results are obtained in the four positions.

- Quasi-static measurements: The same measurement data are used. The deflection $q_2 = \theta_2 - \Phi_2/r_2$ and $q_3 = \theta_3 - \Phi_3/r_3$ is considered. K_2 *and* K_3 are calculated from the relationship:

$$\Delta q_2 = (I_{m2}\ddot{\Phi}_2 - \Delta T_2)\, r_2 \,/ K_2$$
$$\Delta q_3 = (I_{m3}\ddot{\Phi}_3 - \Delta T_3)\, r_3 \,/ K_3$$

Variations with respect to the initial position eliminate the influence of the extra encoder offsets, Figure 3 gives the measured and identified deflection at each joint for a position step of 5° for link 2, resp. link 3.

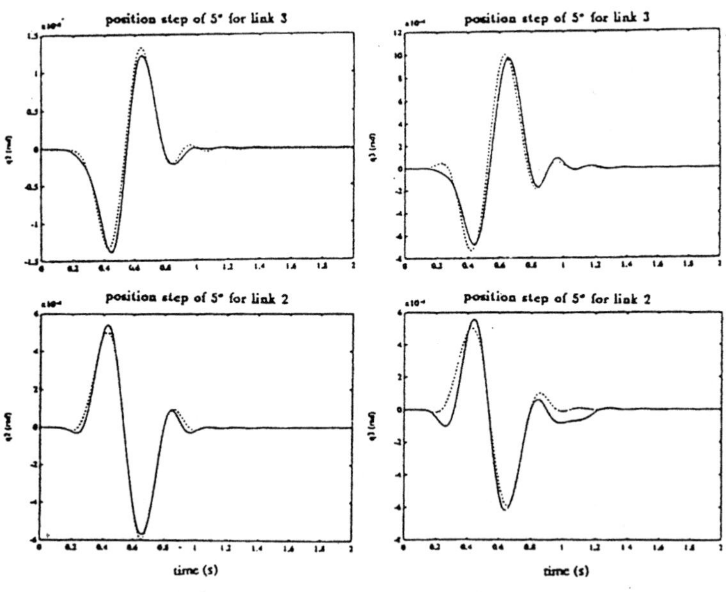

Figure 3
Measured (solid line) and identified (dashed line) deflections at joint 2 and 3 for 5° position steps on links 2 and 3 respectively

- Bandlimited noise measurements: The spring constants are determined from measured data obtained by applying bandlimited noise to the system.

- Stepped sine measurements: Transfer functions are measured using stepped sine measurements. The excitation is generated by the second and third motor separately while the not excited motors are clamped by brakers. The spring constants are determined by comparing the analytical expression for the resonance and anti-resonance frequencies with the measured resonance and anti-resonance frequencies.

In table 2, the identified parameters are compared with the analytical parameters. For control design, the quasi-static values are used. Using these values, the smallest differences between feedforward and actual torques and between measured and calculated state variables are obtained.

		Analytical	Quasi-static	Noise	Static	Stepped sine
A_{22}	kgm^2	597	576	–	–	–
A_c	kgm^2	91.3	115	–	–	–
I_{z3}	kgm^2	214	199	217	–	–
I_{m2}	kgm^2	0.013	0.0158	0.0126	–	–
I_{m3}	kgm^2	0.013	0.0158	0.00899	–	–
$m_3 g x_{3c}$	Nm	835	782	779	–	–
K_2	$10^6 Nm/rad$	–	1.35	1.328	3.183	1.722
K_3	Nm/rad	–	0.841	0.582	1.651	0.626

Table 2: Comparison between the identified and analytical parameters of the model

2.4.3. Control System Design

The second and third axis are controlled with the generalized nonlinear decoupling method explained in chapter 2, section 4. Figure 4 gives the block-diagram of the controller.

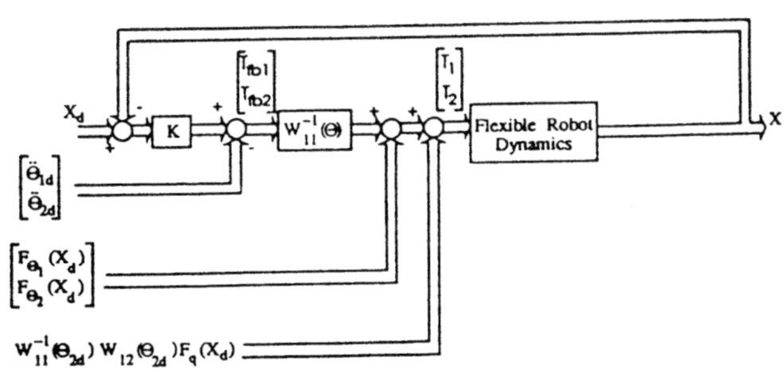

Figure 4: Block diagram of the nonlinear controller for axis 2 and 3

Instead of using a nonlinear state estimator, the velocities are calculated by differentiation. This reduces the number of calculations. To overcome the problem of sentivity to high frequency noise of the differentiator, the measured signals are filtered with digital low pass filters. The filters also reduce the spillover problem, but extends the model order.

2.5. Robot Control

This section deals with control design (control law + trajectory generation) for six axes while previous sections mainly addressed specialised control for one or two axes.

A robot controller consists of two levels: a system level and a servo level. From a specified task (e.g. move from point A to point B on a straight line) the system level generates trajectories for all the joint angles. Two different trajectory generation algorithms have been implemented. The first algorithm is based on a constant acceleration and deceleration profile, and is the standard trajectory generation algorithm for an industrial robot. The second trajectory generation algorithm is based on a 9th order polynomial. It results in a smoother trajectory for the different joint angles.

The output of the system level, i.e. the trajectories for all joint angles, is the input for the servo level. The servo level controls the different motors according to specified algorithms such that the joint angles follow the desired joint angles.

Four different robot controllers with increasing level of complexity are tested:

1. The first robot controller is the standard robot controller that is normally delivered with the robot. It controls all axes in velocity command with an independent PID controller with velocity feedforward, and uses a trajectory based on a trapezoidal velocity profile. This controller is also implemented on the VME-system.
2. The second controller has the same servo level as the first one, but uses a trajectory based on a 9th order polynomial.
3. The third robot controller axis 1 in torque command using the linear position dependent state feedback and feedforward controller described in section 2.2, and axes 2, 3, 4, 5, and 6 in velocity command with an independent PID controller with velocity feedforward, and uses a trajectory based on a 9th order polynomial.
4. The fourth robot controller is used in one test only controls axes 2 and 3 in torque mode with the generalized nonlinear decoupling controller described in section 2.3. The trajectory is based on a 9th order polynomial.

All controllers are optimized (appropriate choice of feedback gains). Integration is added for elimination of steady state errors. A sample period of *6 ms* is used.

The following tests have been executed to evaluate the different robot controllers:

- Pose Stabilisation Time and Pose Overshoot;
- Minimum Positioning Time (time required to travel between two points);
- Test contour for industrial robots: a circle and a straight line.

These tests are further described in chapter 9 of this volume.

2.5. Results

This section discusses the results of the tests performed for the preliminary controller.

2.5.1. Pose Stabilization Time and Pose Overshoot Test Results

Table 3 gives the results of the pose stabilization time and pose overshoot tests. Each measurement is repeated three times and the mean value is reported. The value of the repeatability zone is *0.3 mm*. Points 1 and 2 require and extracting motion of the robot, while points 3 and 4 are near to the robot base and the motion mainly consists of a small rotation around the first axis. Figure 5 shows the results of the extra pose stabilization test for a test motion at extremely high velocities. For controller 1, there is an overshoot of *0.65 mm* and the settling time is *1.60 seconds*. For controller 2, there is an overshoot of *0.16 mm* and the settling time is *0.56 seconds*. For controller 3, there is no overshoot and the end-effector does not leave the repeatibility zone. This corresponds to a null settling time. These results (table 3) lead to the conclusion that a smooth trajectory gives the most important contribution to an improved dynamic behaviour (compare controllers 1 and 2). Only at extreme high velocities, such as in the extra test (figure 5), there is a significant difference between controller 2 and 3. In that case there is still an excitation of the structural dynamics if the controller does not take into account the flexibility.

controller	point 1 → point 2		point 4 → point 5	
	settling time (s)	overshoot (mm)	settling time (s)	overshoot (mm)
controller 1	0.557	2.417	0.300	0.792
controller 2	0.000	0.280	0.000	0.275
controller 3	0.000	0.280	0.000	0.000

Table 3: Results of the pose stabilisation tests

2.5.2. Minimum Positioning Time

The average minimum positioning time for controllers 2 and 3 is *22 %* smaller than for controller 1. These longer positioning times for controller 1 are necessary to avoid excitation of the structural dynamics.

2.5.3. Test Contour for Industrial Robots

The two velocity levels mentioned in this section correspond to *10 %* and *100 %* of the maximum velocity allowed by the standard controller. Figure 6 shows the circle tracking test results of the first three controllers. The bottom figure is a detail of the top figure, taken at the point where the velocity reaches its maximum. The solid line is the result obtained at low velocity, and the dashed line is the result obtaines at high velocity. The maximum deviation for controller 1 is *1.5 mm*, for controller 2 *0.7 mm*, and for controller 3 *0.5 mm*. The straight line tracking test results of controllers 1, 2 and 3 are the same as for the circle tracking tests: controller 1 is the worst, and controllers 2 and 3 are the best. Controller 3 is slightly better than controller 2.

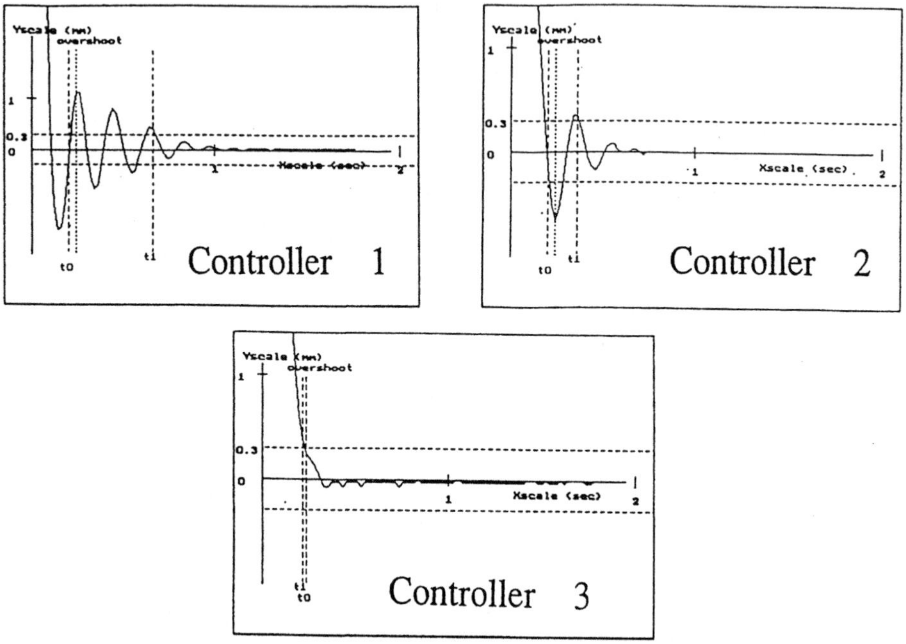

Figure 5
End effector response for a test motion at extremely high velocity with controllers 1, 2 and 3

Figure 7 shows the straight line tracking test results of controller 2 and 4. Controller 4 compensates for static errors due to the gravitation. Figure 8 compares the dynamic tip position errors normal and tangential to the desired trajectory for controller 2 and 4. The maximum dynamic tracking error normal to the desired trajectory reduces from *1.6 mm* to *0.8 mm*. The maximum dynamic tracking error tangential to the desired trajectory reduces from *2.5 mm* to *2.0 mm*.

2.6. Conclusion

The tests have shown that the performance of industrial robots can be improved with modifications of the industrial robot controller. The change from a trajectory based on a trapezoidal velocity profile to a smooth trajectory gives the largest contribution to the improved dynamic performance, but there is a significant improvement of a flexible controller over a classical PID controller in terms of tracking accuracy. The tests have also shown that a nonlinear flexible controller for axes 2 and 3 compensates the static deflection and improves the dynamic accuracy. These improvements are small due to the limited flexibility of joints 2 and 3.

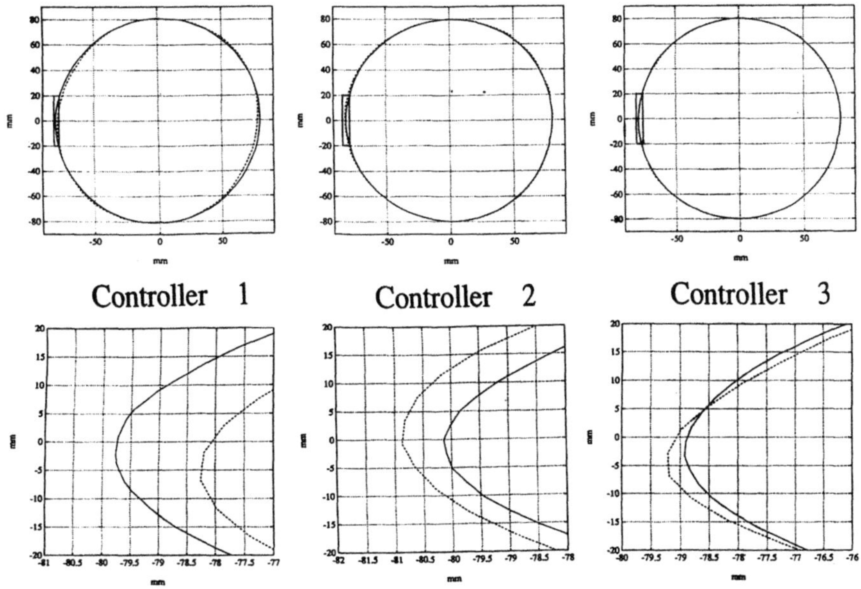

Figure 6
End effector response measured during tracking tests with controllers 1, 2 and 3. solid line: low velocity; dashed line: high velocity

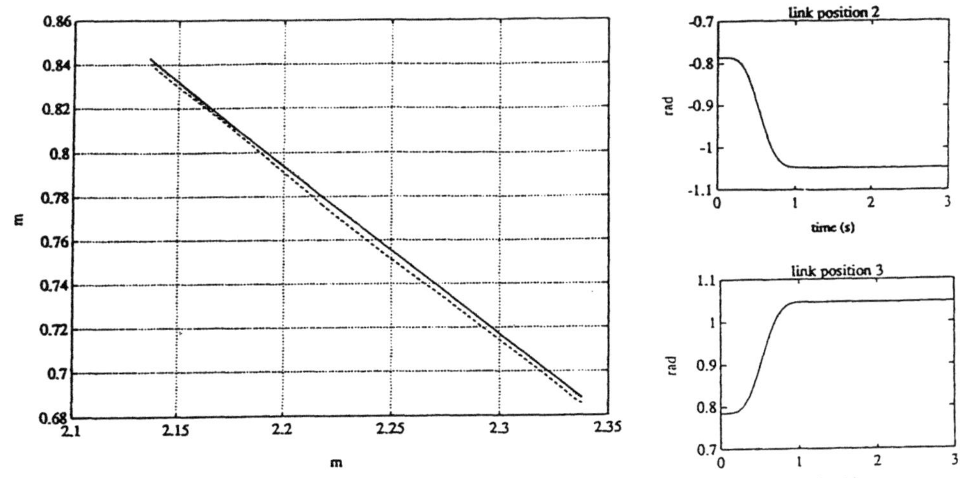

Figure 7
The tip position trajectory and link position trajectories of a generalised nonlinear decoupling controller (dotted line) and a rigid controller (dashed line) compared with the desired trajectory (solid line)

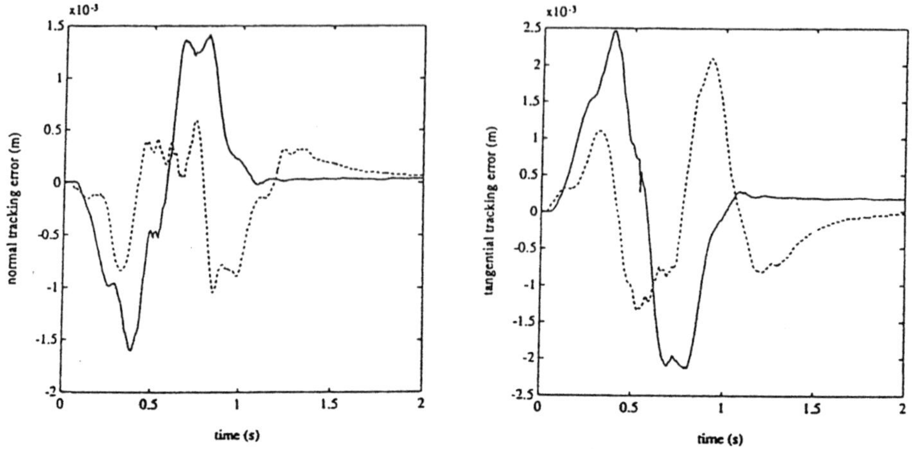

Figure 8
Dynamic tracking error normal and tangential to the desired trajectory in function of time for a rigid controller (solid line) and a flexible controller (dashed line)

3. Implementation of an Antivibration Control in an Industrial Robot Controller Prototype

3.1. Industrial Constraints and Design Trade-Offs

The goal of the industrial demonstrator was to prove the compliance of the antivibration robot control algorithms with general robot control specifications and implementation constraints implied by industrial applicability considerations.

The specifications to be met by the robot control were, as for the preliminary developments carried out at K.U.Leuven:

- a short positioning time,
- a complete damping of positioning overshoots and vibrations,
- a small trajectory tracking error,
- robustness with respect to varying robot configurations and unmodelled vibration modes.

Nevertheless, in this final task of integration of the antivibration control algorithms in the industrial demonstrator, additional industrial constraints had moreover to be complied with, while maintaning the level of performance reached by the K.U.Leuven researchers on their Laboratory VME-based multiprocessor.

The target hardware on which the new software has been implemented is the high performance servo level board LAR 830/286 developed by AEG AG in the framework of SACODY and interfaced with the AEG robot controller IRC 500 System Level environment (see Chapter 7). The LAR 830/286 board, in charge of the servo-level software is also interfaced with the new Input/Output and sensor signal processing board (LAR 831/186).

The software to be developed had therefore to cope with the characteristics of this implementation environment that were:

- compatibility with a conventional system level software generating trapezoïdal-velocity trajectory profiles
- synchronisation with the Input/Output hardware which led to design a servo control running with a *6 ms* sampling period and robust with respect to a *6 ms* control order delay.
- a 80286-based servo-level hardware implying a fixed arithmetic based programmation.

3.1.1. Compatibility with Trapezoidal Velocity Trajectory Profiles

The velocity profile generated within the System Level of the industrial prototype is trapezoïdal. At the end of each *48 ms* System Level sampling period, a new position order is sent to the Servo Level. These positions are calculated according to the nominal robot dynamic parameters specified by KUKA, i.e: the maximum velocities and the acceleration times.

In order to implement a trajectory smoothing similar to that proposed and tested by K.U.LEUVEN, the original trajectory coming from the System Level was interpolated from *48 ms* to *6 ms* and then filtered.

Figures 9 and 10 compare the different acceleration and velocity profiles tested during the development:

- *The ideal profile* as calculated in the System Level,
- *the interpolated profile* which is the ideal profile after interpolation from *48 ms* to *6 ms*,
- *the filtered profile* which is the interpolated profile filtered by a first order filter,
- *the K.U.LEUVEN profile* which is a smooth polynominal profile directly sampled at the servo level rate from the desired path.

The interpolation and smoothing process of the industrial implementation architecture generates some *tracking error* that needs to be minimised. A *48 ms* time constant for the first order filter provides a good compromise between this tracking error and the smoothing of the trajectory:

- the cut off frequency is *3.3 Hz* whereas the lowest vibration mode of the robot IR160 has been measured at *5.1 Hz*.
- the time constant of the first order filter, equalling the system level sampling period (*48 ms*), leads to a maximum tracking error equal to the maximum difference between to consecutive points generated by the System Level.

3.1.2. Synchronisation with the Input/Output Hardware

The controller I/O board LAR 831 has a maximum sampling rate about *2 ms*. Nevertheless the minimum delay between sensor readings and the release of a control order with a loop closure in the LAR 830 is about *3 ms*.

In order to take into account the calculation time of the new control law, and to limit the delay between input and output to the servo sampling period, a *6 ms* LAR 830 sampling period and a *3 ms* LAR 831 sampling period have been chosen.

The diagram figure 11 describes the overall structure of the servo level software:

- The initialisation box is only activated by a General Reset.
- The background box prepares the data for the servo-loop, gives informations back to the system level, etc...
- The servo-loop box performs the computation of the control loop each 6ms period and activates the background box at the end of each 48 ms period.
- The I/O timer, programmed to work with a *3 ms* cycle, activates the I/O board LAR 831. Every two timer cycle, the servo loop is also activated, and therefore works with a *6 ms* cycle time.

3.1.3. Programmation Constraints

The target hardware LAR 830 features an INTEL 80286-10Mhz processor but no coprocessor.

The ratio between integer operations and floating point operations is about 100 (and should be reduced to 10 with a 287 coprocessor). This led to an implementation of all the algorithms with integer variables and operations. To keep a good accuracy in the calculations the variables were coded with *32-bit* integer and the gains with *16-bit* integers.

An in-depth analysis of the algorithms has been carried out in order to avoid any overflow or underflow of variables, especially concerning intermediate results.

3.2. Servo-Loops Structures

The servo loop structure has been designed in 3 steps:

The first step was a redesign of the servo-loop but keeping a conventional control structure that is a Proportional and Integral controller with a velocity feedforward. In this configuration, *the output of the digital controller is a velocity command*, and the velocity loop is closed in the power amplifier through an analog corrector.

The second step was the implementation of digital velocity control loops. The velocity loops were closed either from the time derivative of the motors position read by encoders or from the tachogenerators signals via analog to digital converters. In this configuration, *the output of the digital controller is a torque command*. In this step, anti-spillover filters were introduced for the velocity estimations and acceleration feedforwards were added in order to improve the tracking capabilities of the control law.

The third step was the implementation of the flexible modes control law using the extra encoders information. In this case the controller is a 4th-order discrete-time state-feedback controller, with an acceleration feedforward to improve the tracking error and an integral action to improve the steady-state accuracy. An estimator calculates velocities by differentiation of the encoders signals and filters high frequency noise. The proposed control scheme is further discussed next section.

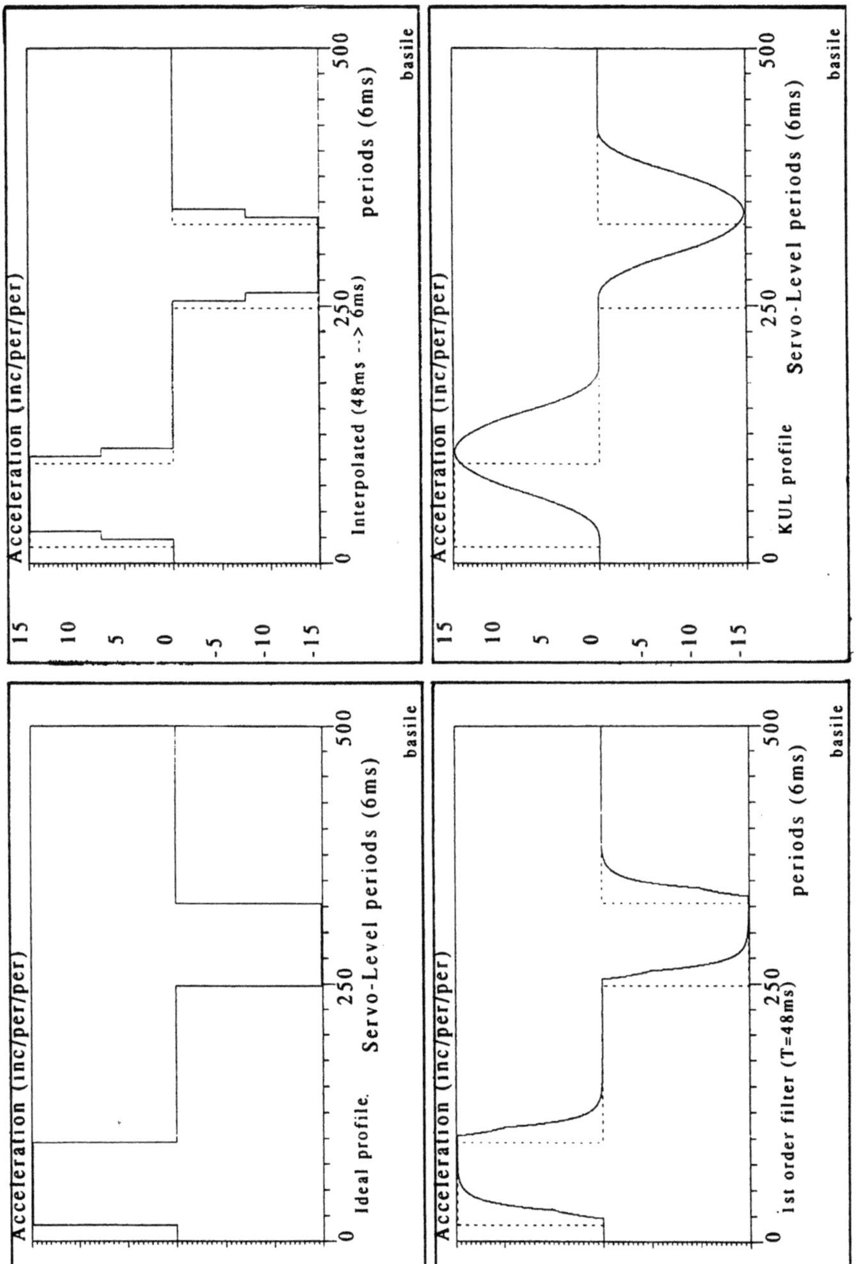

Figure 9: Acceleration Profiles

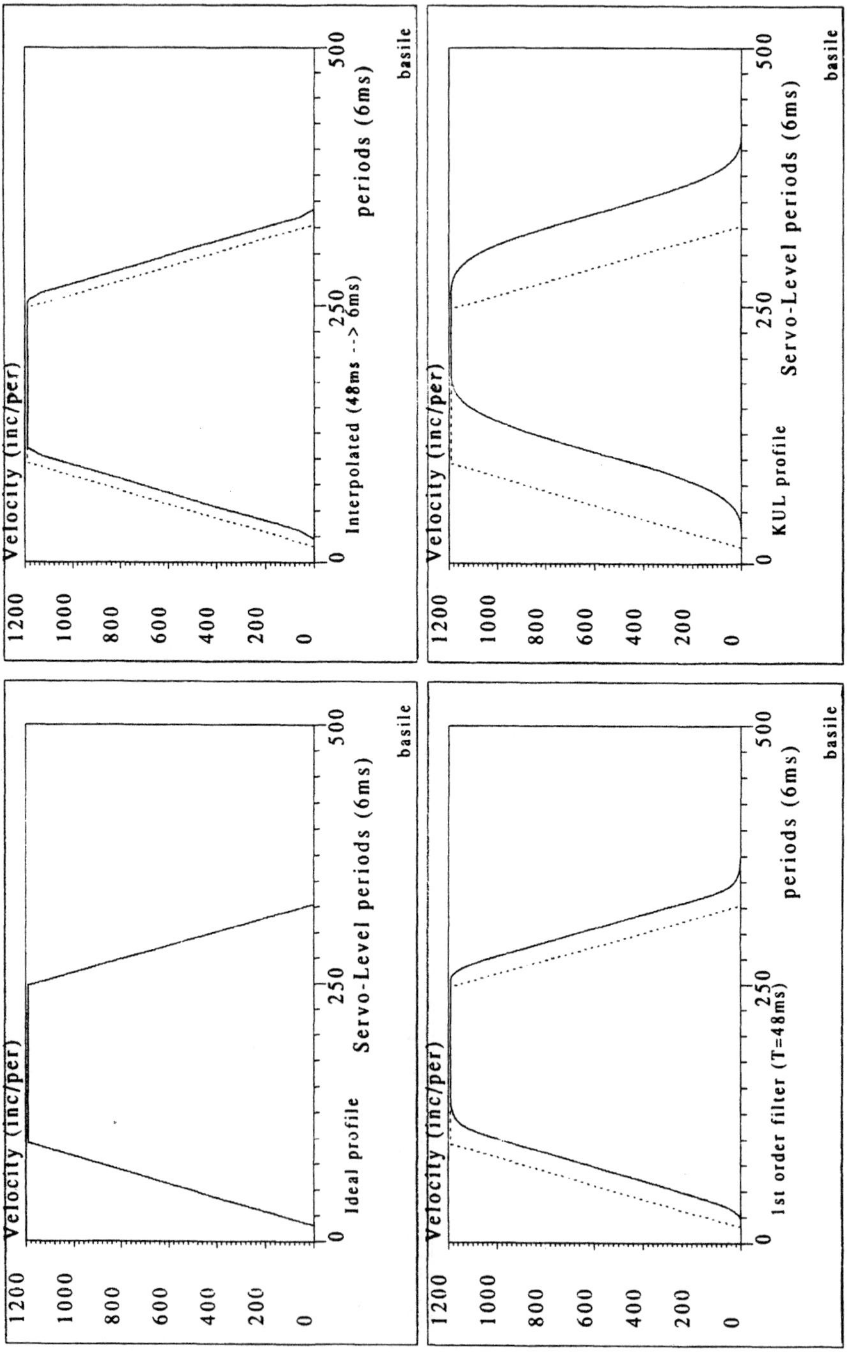

Figure 10: Velocity Profiles

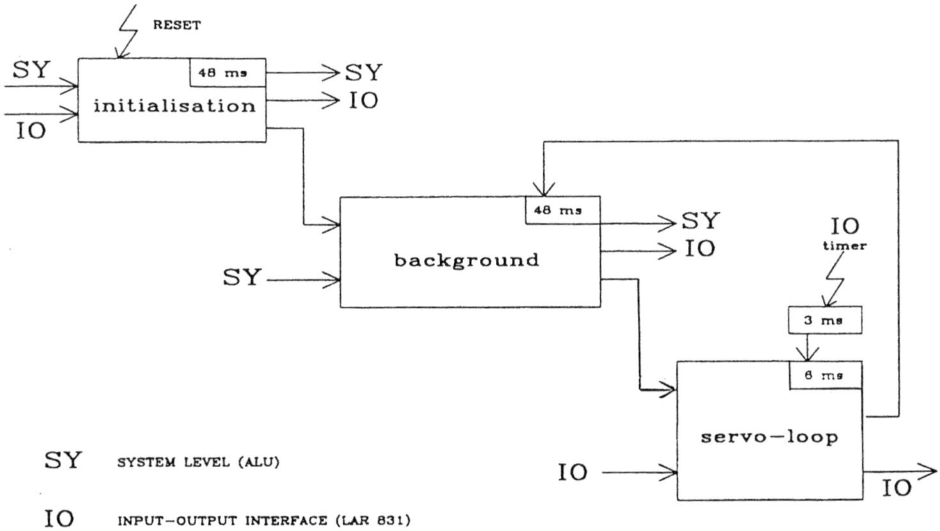

Figure 11: Servo loop software overall diagram

3.3. Antivibration Control Scheme

As it has been developed in the previous sections, a control architecture based on four different control tunings had been successfully implemented at K.U.Leuven on a laboratory multiprocessor. The advantage of such a structure was a control remaining optimal, whatever the robot load and extension conditions.

In the case of the AEG industrial controller prototype, it was not possible to calculate the output of two different servo loops and then to interpolate their output, due to the global calculation load of the LAR 830. It should have been possible to interpolate the control loop parameters at the system level sampling period rate. But stability and robustness problems could then have arisen. Finally an intermediate and fixed-tuning solution, engineered in simulation by the implementation of control robustness methods, has been chosen for the first axis. As it is described in chapter 9, this control was still shown to give much better results than the conventional control schemes.

Figure 12 describes the overall control scheme that was implemented by BERTIN within the AEG industrial controller prototype to control the robot KUKA IR 160: Only the two first main axis are controlled with the new control law. The third axis and the wrist axes are still controlled using a conventional control scheme. The velocity of the first axis is measured by the tachogenerator signal via an analog to digital converter included in the LAR 831. The velocity of the second axis is as for it estimated by the differentiation of the motor encoder data.

It appears from the result that the active damping of the first two modes of the KUKA IR 160 robot yields all the expected improvement. It means that the natural damping of the other vibration modes is sufficient and that these residual modes are not so excited as they were before, probably due to the non-linearities encountered with large vibrations on the first mode.

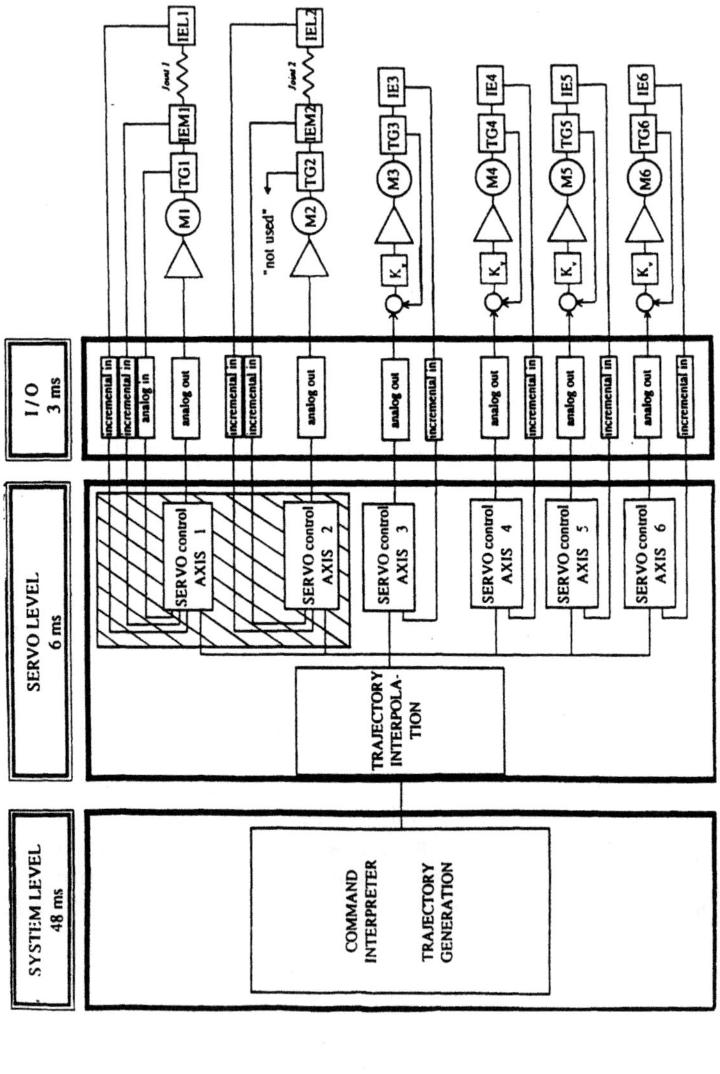

Figure 12: Overall Control Scheme

TG : Tacho Generator
IE : Incremental Encoder
IEM : Incremental Encoder on Motor side
IEL : Incremental Encoder on Load side of the joint
K_v : Analog velocity control loop

Bibliography

[1] Gerung, M.: Accuracy measurement on a KUKA IR 161/60 with an RCM3. Technical Report, Esprit project 1561: SACODY, 1990.

[2] Van Der Auweraer, H.: Off-line identification of a KUKA IR 161/60 industrial robot. Technical report, task 4851, Esprit project 1561: SACODY, 1990.

[3] Swevers,J., De Moor, B., Van Brussel, H.: Stepped sine system identification, errors-in-variables, and the quotient singular value decomposition. Proc. of the 15th International Seminar on Modal Analysis, Leuven , Belgium, pp. 59-76, 19-21 September 1990, and Mechanical Systems and Signal Processing Vol. 5, No. 6, November 1991.

[4] Van Den Bossche, J.: RODYM: A new approach to robot metrology. Internal report, K.U. Leuven, Department of Mechanical Engineering, Leuven, Belgium, 1990.

Chapter 9

Robot Performance Testing

Heinz Wörn, Hans-Richard Tradt, Michael Gerung
KUKA Roboter GmbH

1. Introduction

This report contents the documentation and the analysis of the accuracy measurements performed on a KUKA IR 160/60 industrial robot with two conventional robot controllers, that will be referred to in the sequel as the RC1 and the RC2 controllers, and with the newly designed controller RC3 developed during the project SACODY and performing the antivibration control of the robot (see Chapter 8).

The RC1 is the standard robot controller interfaced with the KUKA IR160/60. The RC2 and the RC3 implement the same numerical controller hardware, developed by AEG and presented Chapter 7 in this volume. They differ by the control algorithms they implement, RC2 implementing a conventional control architecture and RC3 the new SACODY control concepts.

The tests presented and compared in this chapter have been performed in order to give information about the improvements in accuracy and cycle time of industrial robots brought by the new anti vibration controller.

2. Scope

The criteria for the different controllers tested refer to the accuracy of the Industrial Robot in terms of positioning and path following and to the robot total positioning time which includes the minimum positioning time and the robot settling time.

To investigate these characteristics in an accurate and efficient way, five different tests procedures have been prepared. They are partly extracted from the existing robot test standards of the International Standard Organisation (ISO 9283) and of the Robotic Industries Association (ANSI/RIA R15.05) and partly based on internal KUKA tests. The measurements implemented by these procedures concern:

- The Pose Settling Time and The Pose Overshoot, that characterise the time taken by the robot tool to stabilise at the desired pose,
- the Minimum Positioning Time, that is the time required for the movement between two different poses,
- the Cornering Deviations (registration of the path accuracy),
- the RIA standard test path (registration of the cycle time and path accuracy),

- the KUKA test contour for Industrial Robots, that allows qualitative and quantitative criticism of the robot system path-following performance.

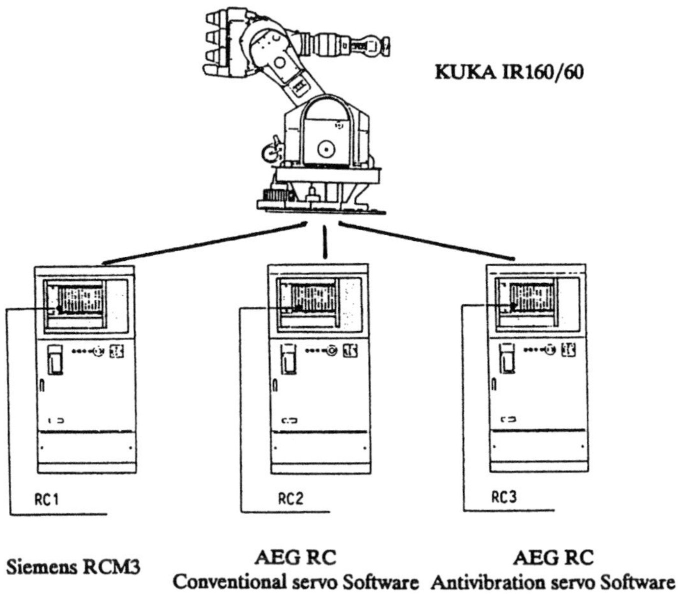

Figure 1: The tested robot systems

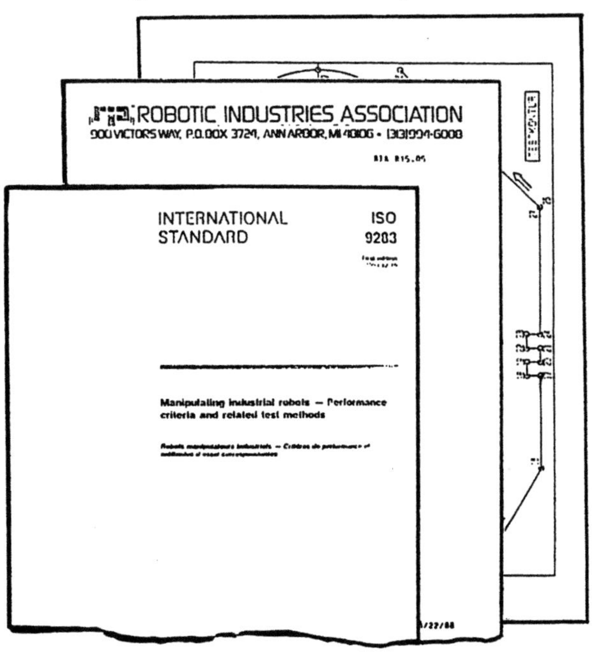

Figure 2: The applied procedures

3. Description of the Test Conditions and Test Sequences

3.1. The ISO Test Conditions

3.1.1. Measured Positions

The specification ISO 9283 defines five robot positions to be measured. The five corresponding robot end points are located in a plane which is placed in a cube in the working space of the robot. The cube has to fulfil the following requirements:

- The cube shall be located in the portion of the working space with the greatest anticipated use,
- the cube shall have the maximum volume allowable with the edges parallel to the base coordinate system.

The test plane reaches from the lower edge facing towards the robot to the upper edge of the opposite cube plane. The ISO point P1 is the intersection of the plane diagonals and corresponds to the center of the cube. The four other ISO points P2 to P5 are located at a distance from the ends of the diagonals equal to *10 %* of the diagonal length. The following figure illustrates the position of the ISO-Points.

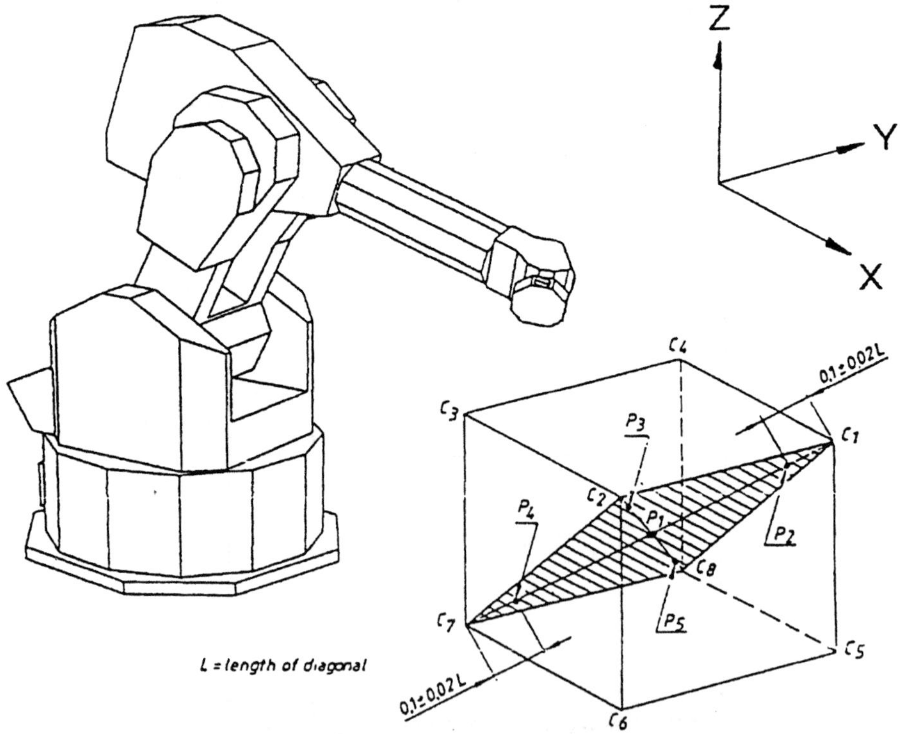

Figure 3: Location of the Measured ISO Positions

3.1.2. Test Velocities

All the pose characteristics must be tested at the maximum velocity achievable between the specified poses, i. e. the speed override set to *100 %*.

For path characteristics, the tests have to be conducted at *100 %*, *50 %* and *10 %* of the rated path velocity, equal to *100 m/min* for the KUKA IR160/60.

3.1.3. Test Load

All the tests must be executed at the maximum rated load conditions (mass, gravity center position, moments of inertia) according to the manufacturer's specification.

The rated load as specified by KUKA for their IR160/60 with a *400 mm* arm extension is *45 kg*. The welding gun which was used as a tool during the measurements performed during the SACODY project weighs *46.5 kg*. The specified position of the mass center is shown on Figure 4.

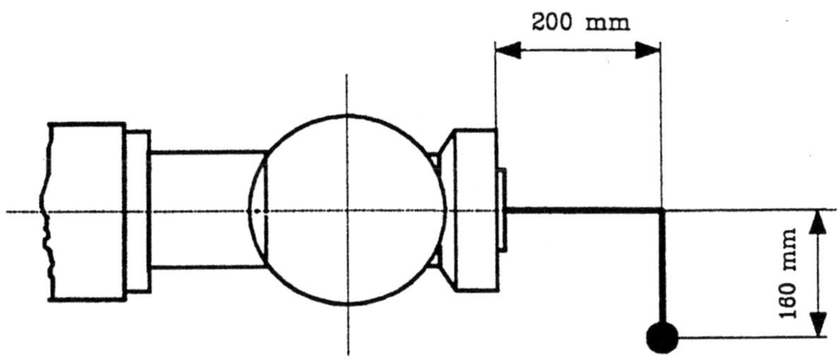

Figure 4: Position of the Load Center of Gravity

3.1.4. Test Acceleration

There is no certain acceleration specified in the ISO Test Specification 9283. Therefore, the acceleration that has been used in the tests is *100 %* of the acceleration specified by the manufacturer.

3.2. The Pose Settling Time and Pose Overshoot

According to the ISO 9283 the pose stabilisation time is the period of time which elapses between the instant when the robot gives the attained pose signal and the instant when the oscillatory motion or the oscillatory motion of the mechanical interface falls within a specified limit. This stabilisation time is also referred to in the RIA standard as the "Settling Time".

In case of a damped motion of the mechanical interface, the overshoot is defined as the distance between the actual position at the controller "attained pose" signal and the nominal position (the overshoot is then negative).

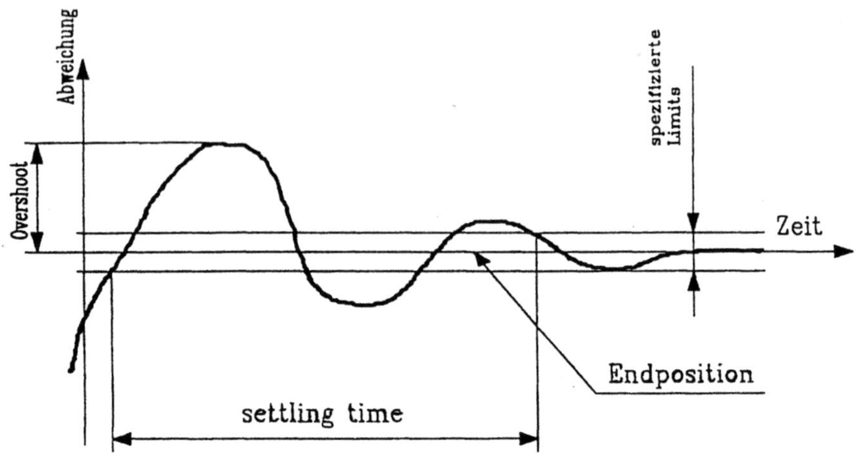

Figure 5: Definition of Pose Settling Time and Pose Overshoot according to RIA - Standard

The different test standards used during the project all define the stabilisation times with respect to the *attained position signal*, that is a controller-dependent signal. Since different controllers of different manufacturers (SIEMENS and AEG) had to be compared, it has been preferred to measure the stabilisation time with respect to an external event. With the sensor system RODYM, presented chapter 6 in this volume, the time elapsed between the instant of the first entrance of the probe into the specified limit band and the moment when the robot settles within this specified limit band is retained as a measurement for the settling time (according to the RIA Standard). In the case of a damped motion, the system RODYM gives an overshoot of *0 mm*.

3.3. Minimum Positioning Time

The minimum positioning time is the time between departure from and arrival at a stationing state (signal "position reached") when traversing a predetermined distance under pose - to - pose control. The measured points (distances) are defined as shown on figure 6.

The points are taught along the main diagonal of the cube (Figure 1) between the points P2 and P4. For each travel the mean value of three cycles is calculated.

The time taken by the system RODYM is however the time between leaving and entering the region of specified limits (*+/- 0.3 mm*) of the two positions.

3.4. Cornering Deviations

Cornering overshoot and *cornering round-off* are the deviations between a command path and the actual path when the command path consists of two lines perpendicular to each other.

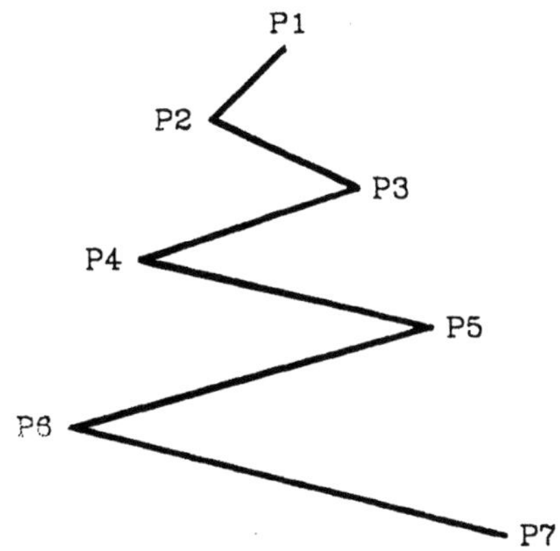

Section	Distance [mm]
P1 – P2	10
P2 – P3	20
P3 – P4	50
P4 – P5	100
P5 – P6	200
P6 – P7	500

Figure 6: Definition of points and illustration of the cycle for measurement "Minimum positioning time"

Cornering round-off error is defined as the minimum distance between the corner point and the attained path. Cornering overshoot is defined as the maximum deviation from the command path after the robot has started on the second path without time delay and with programmed constant path velocity. The start position is halfway between P5 and P2. From there the path P2 - P3 - P4 - P5 will be travelled in CP-mode.

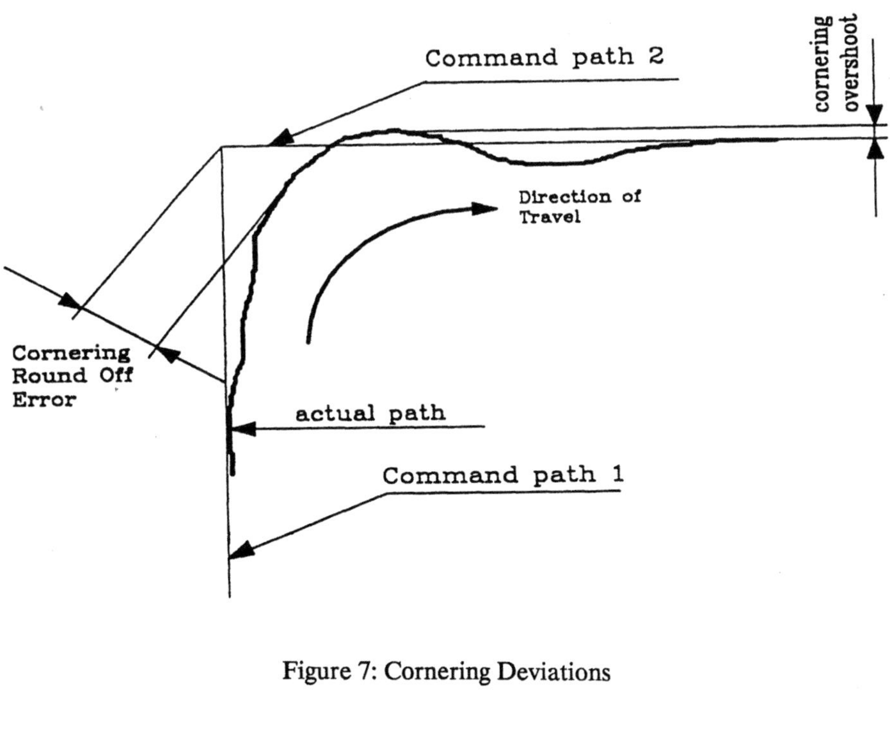

Figure 7: Cornering Deviations

3.5. RIA Standard Test Path

3.5.1. Measured Positions

The standard-test-path (STP) is located in a plane parallel to the (1, -1, - 1) plane of the robot base coordinate system. This plane (standard-test-plane) is positioned so that the center of the robot work space *(c_w)* lays in it. The center point c_w is the midpoint of the line, which is parallel to the base *x*-axis, whose *z*-axis position is the point of maximum horizontal travel and whose endpoints are the points of intersection with the robot working space. Through this point passes the center line of the STP. For the SACODY Project, only the cycle time test of the RIA Test specification has been implemented.

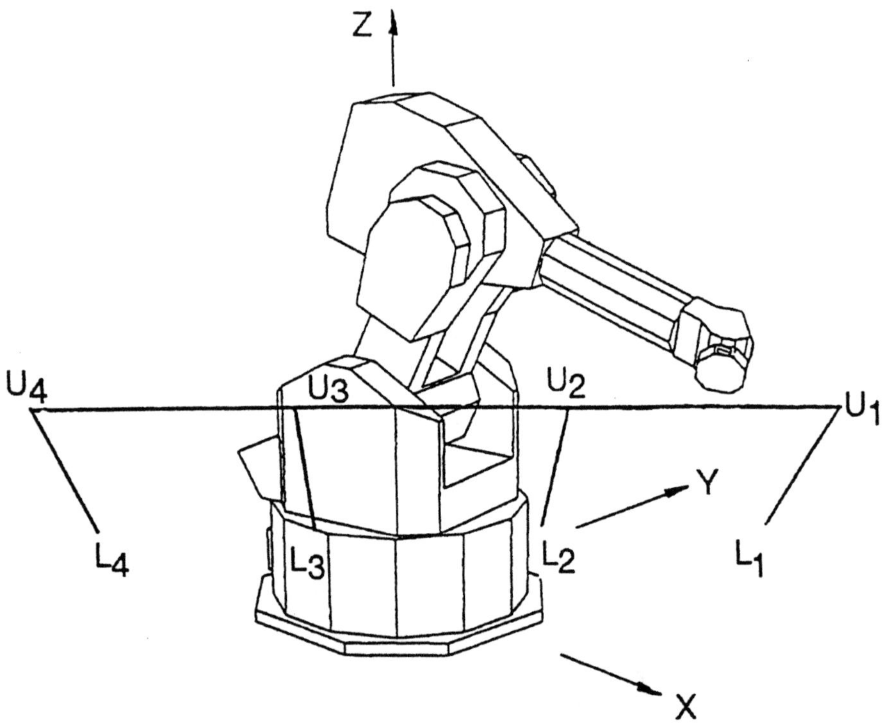

Figure 8: Location of the RIA Standard Test Path

3.5.2. Cycle Time Measurement

Two figures of merit are reported under cycle time: "*the segment cycle time*" and "*the traverse speed*". The segment cycle time is a measure of the average time required by the robot to travel through one segment of the Standard Test Path. Each segment motion consists of an upward movement, a horizontal movement and a downward movement.

The segment cycle time for segment one is the elapsed time between departure from L_1 until arrival at L_2 when travelling through U_1 and U_2. Analog to this is the definition of the cycle time for segments 2 and 3.

The segment cycle time is reported as the average number of seconds required to move through one segment of the Standard Test Path.

The traverse speed is the average speed attained by the robot while making relatively large movements. It is measured during the latter portion of the cycle time test when the robot is traversing the entire length of the Standard Test Path.

The traverse speed shall be reported as the average linear speed achieved in moving from point L_4 (through U_4, U_3, U_2 and U_1) to point L_1 on the return portion of the test cycle.

3.6. Test Contour for Industrial Robots

The test contour is travelled through with different velocities and also in fly-by-mode. The results serves more for qualitative than quantitative criticism of the path characteristics of an industrial robot. To enable a comparison of the certain measurements, the path accuracy is measured in some significant parts of the contour and calculated. These parts are marked in Figure 7 (Pos. 1 - Pos. 3).

The first measurement with a velocity of *25 m/min* and high-accuracy-mode (fly-by-factor *0 %*) serves as reference path.

The plane of the contour lays parallel to the base X-Y-plane and contents ISO - point P1. Relative to the contour coordinate system point P1 has the coordinates (*[300/0/0]* (see Figure 9).

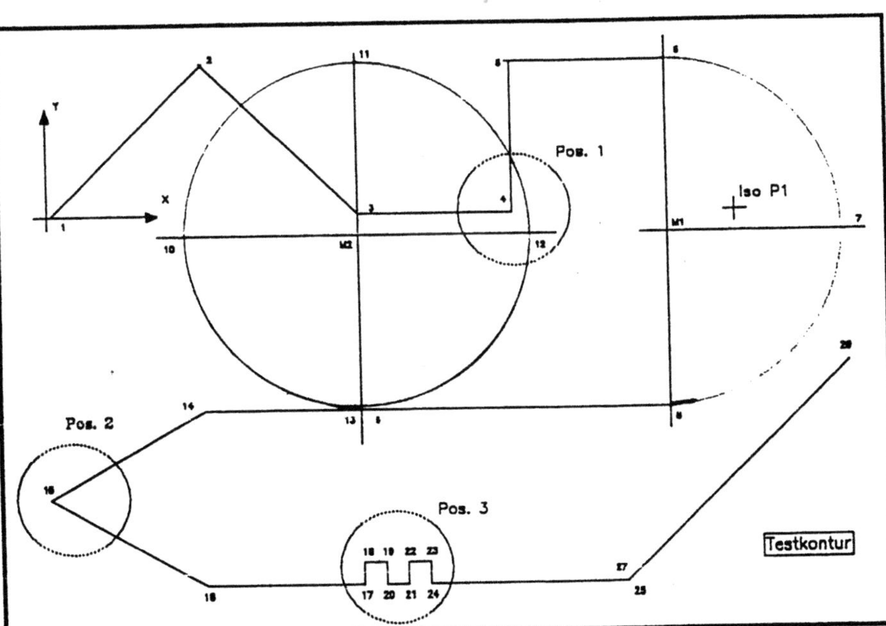

Figure 9: Measured Positions of the Test Contour

4. Performance of Measurements and Test Results

4.1. Measurement Components

Measured Object : KUKA IR 161/60 Armextension AV400
Ser.-Nr. 3024
Handling-weight *46.5 kg* (welding-gun)

IR-controllers : RC1 Standard KUKA controller
RC2 Sacody controller with conventional control (velocity command)
RC3 Sacody controller with antivibration control

Measurement system : RODYM-system of Krypton-Metrology with 2D digitizing tablet 350 * 450, measurement probe, printer

Further devices : Oscilloscope with digital memory

4.2. Point-To-Point Measurements

4.2.1. Settling Time and Pose Overshoot

In this test, the results of which are shown Figure 10 only the ISO-positions P2 (example for a extended Position) and P5 (example for a folded position) are measured. The measurement is triggered with a position window on the RODYM board. For this, a region around the programmed position (*+/- 5 mm* in *X* - and *Y* - direction of the board) is defined; as soon as the probe enters this region, datas will be collected for *2 s*. The specified limit is *+/- 0.3 mm* around the nominal position. The direction of travel is from P1 to P2 and from P4 to P5. Each point is measured three times and as result the mean value is calculated.

With velocity command there was measured only an extremely extended position of the robot with only robot axis one moving. The reason for this is, that in this position the effect of the new controlling algorithms is most visible.

Measurement	Position	RC1	RC2	RC3
Settling Time (sec)	Extended		0.890	0.220
	P2	0.557	0.495	
	P5	0.300	0.100	
Pose Overshoot (mm)	Extended		9.320	1.200
	P2	2.417	2.938	
	P5	0.792	0.482	

Table 1: Settling Time and Pose Overshoot

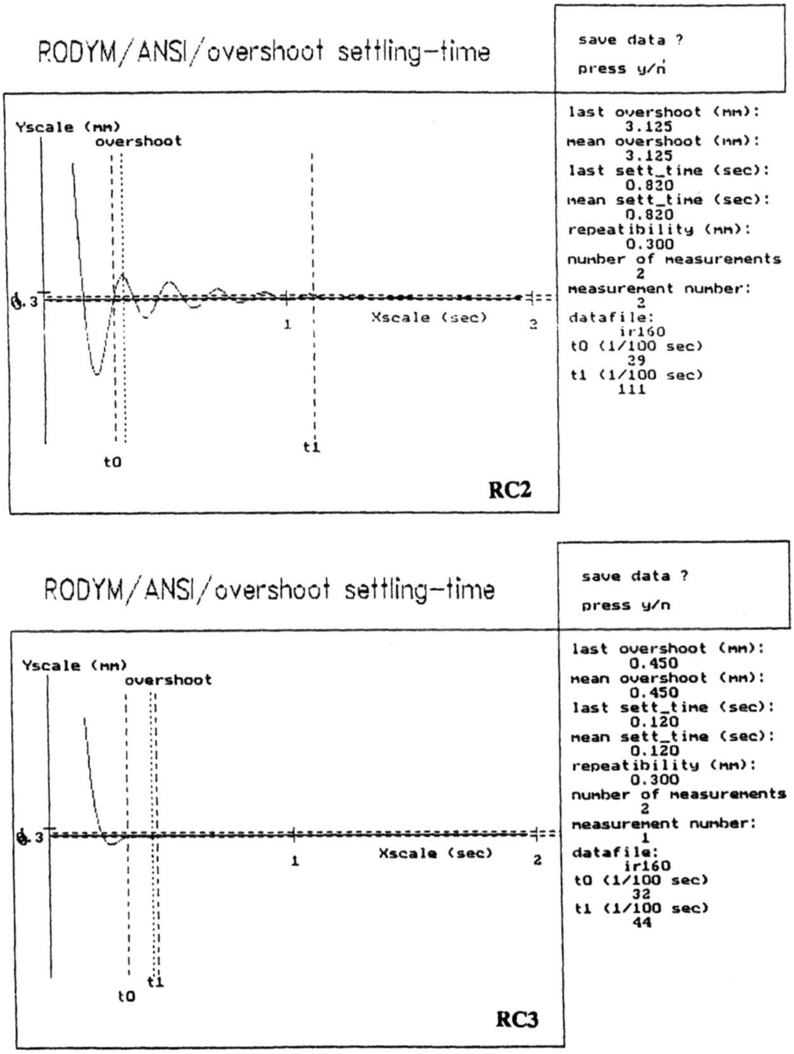

Figure 10: Measurement Plots for Settling Time and Pose Overshoot

4.2.2. RIA Standard Test Path

For the KUKA IR 161/60 the Standard Test Path consists of three segments with the following distances:

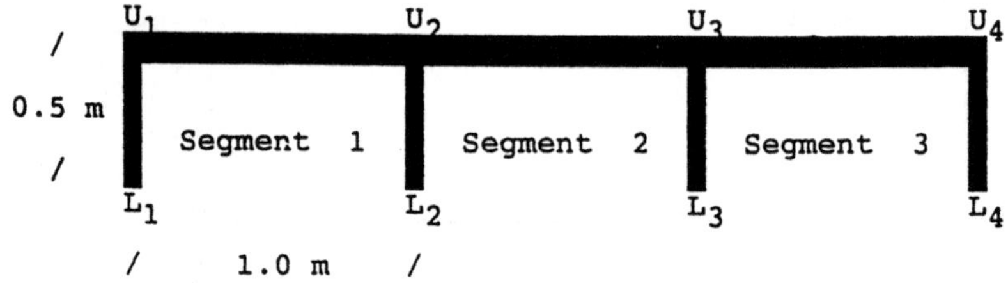

Figure 11: Segment of the RIA Standard Test Path

The cycle time is measured with an output signal of the controller. The output is reset just before the movement begins and set again when the robot arrives at the measured position. The cycle time is the time between the falling slope of the first signal and the rising slope of the following signal.

Measurement	RC1	RC2	RC3
Mean Segment Cycle Time	2.53 s	2.20 s	2.19 s
Traverse Speed	0.94 m/s	1.50 m/s	1.46 m/s

Table 2: RIA Standard Test Path

4.3. CP Measurements

4.3.1. Minimum Positioning Time

After the ISO - Specification this measurement should be performed in PTP - mode. Since it is not possible to reach the defined points in PTP - movement without touching or even damaging the Rodym - board, the movement is programmed in CP - mode.

Travelled Distance (mm)	RC1	RC2	RC3
10	0.160 s	0.129 s	0.227 s
20	0.200 s	0.288 s	0.318 s
50	0.403 s	0.507 s	0.477 s
100	0.597 s	0.659 s	0.696 s
200	0.900 s	0.999 s	0.999 s
500	1.500 s	1.544 s	1.612 s

Table 3: Minimum Positioning Time

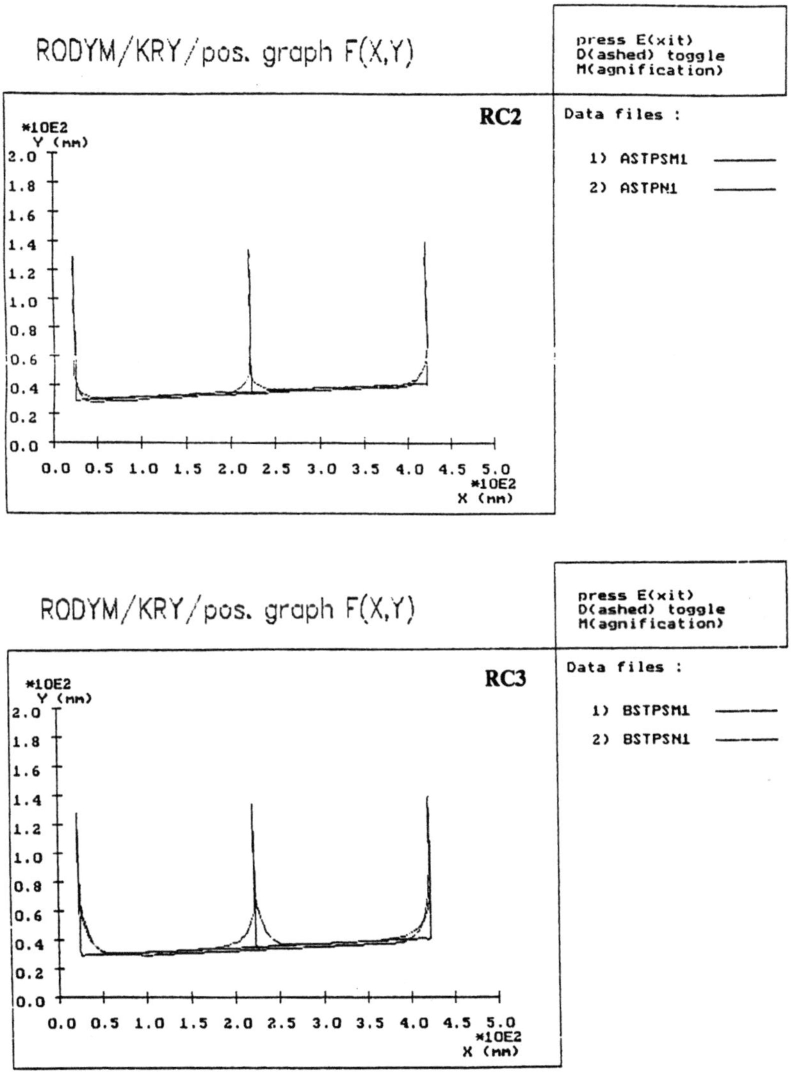

Figure 12: Measurement Plots for RIA Test Path

4.3.2. Cornering Deviations

The whole path defined in ISO 9283 has not been measured, but only a part before and after each corner point (active area of the RODYM measuring board). The measurement was triggered when the probe entered a predetermined region (+/- *5 mm*) at the edge of the measuring range of the board along command path 1 (see Figure 5).

The first measurement is taken in high-accuracy mode and a velocity of *100 m/min*. With this values a reference path is determined. The path then is measured with fly-by-mode.

The fly-by-mode is different in the controllers:

- RC1: no reduction of velocity around the corner points (it was not possible to travel around the corners with full speed so the velocity was reduced to *25 mm/min*).
- RC2/RC3: a certain distance to the corner point is defined with which the controller calculates a path around the corner point (the distance was set to *25 mm and 50 mm* with a velocity of 25 m/min).

Measurement	Position	RC1	RC2	RC3
Cornering Overshoot (mm)	P2	0.00	6.4	0.8
	P3	1.83	-	-
	P4	0.33	3.5	-
	P5	3.52	-	-
Cornering Round-Off Error (mm)	P2	4.34	27.3	16.7
	P3	2.55	-	-
	P4	4.97	18.7	-
	P5	4.68	-	-

Table 4: Cornering Deviations

4.3.3. Test Contour for Industrial Robots

For the measured paths in high-accuracy-mode, there were only visible differences in the path accuracy at the circles (see Figure 9). Pos. 1 means the biggest deviation between the measured circles; the measured position must not be necessarily at the place shown in Figure 9.

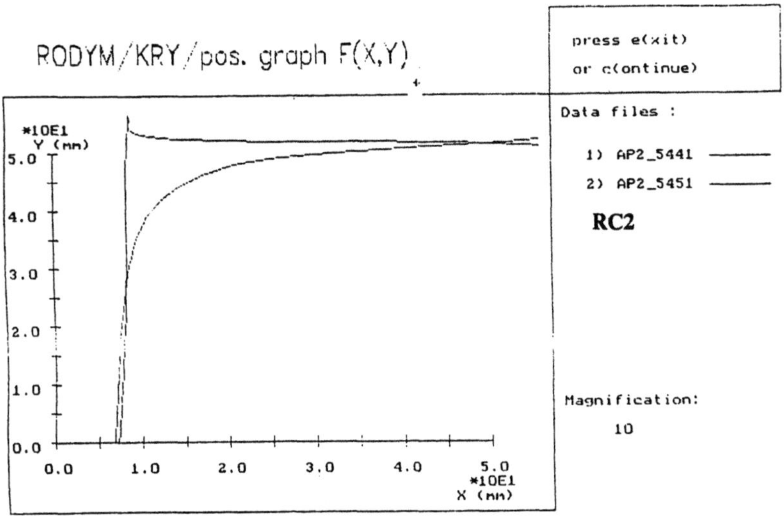

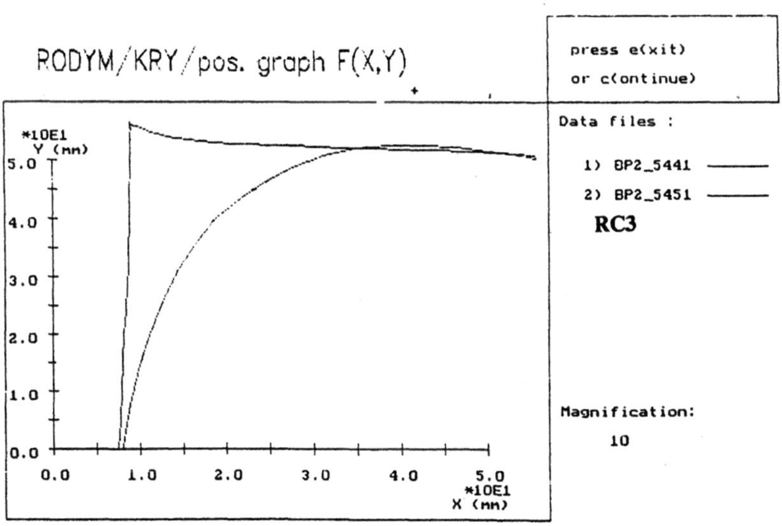

Figure 13: Measurement Plots for Cornering Deviations

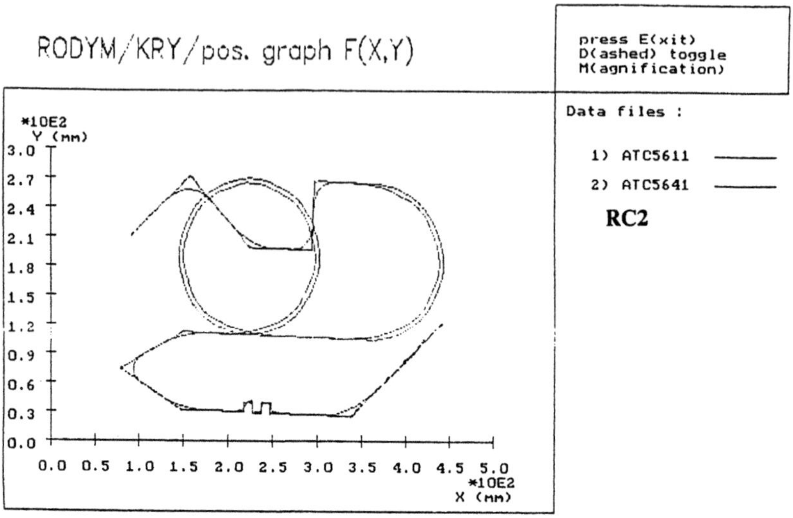

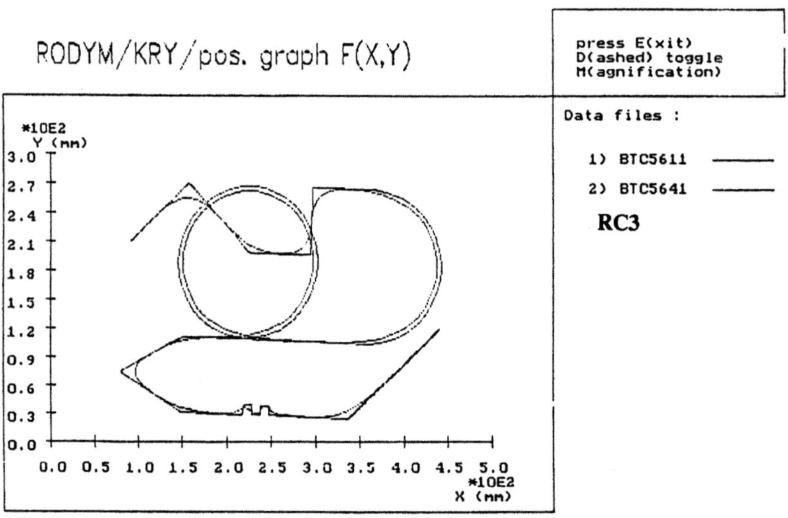

Figure 14: Measurement Plots for Test Contour for Industrial Robots

The fly - by - distance for RC2 and RC3 were set to *50 mm*. In RC1 the fly - by factor was set to *100 %*.

Velocity	Position	RC1	RC2	RC3
50 m/min (High Accuracy Mode)	P1	1.5	4.2	15.0
100 m/min (High Accuracy Mode)	P1	1.5	5.5	15.0
25 m/min (Fly-By Mode)	P1 P3 P4	0.0 3.9 3.4	5.7 16.2 7.0	7.9 17.8 4.9

Table 5: Test Contour

5. Discussion of the Test Results

5.1. Accuracy

For this performance criterion the results for cornering deviations (see 4.3.2) and the test contour (see 4.3.3) serve for a classification.

The cornering overshoot with RC3 is eight times smaller than with RC2. The overshoot with RC1 is zero. Really comparable are RC2 and RC3 because both of them use the same fly-by mode. Also the cornering round off error is *40 %* smaller with RC3 than with RC2.

The results for the test contour show that the deviation of the path in the circle is in all cases bigger with RC3 than with RC2 and RC1. The only smaller distance in the test path occurs at position 4 (see Figure 9); there the deviation is *30 %* smaller than with RC2.

The CP - results show that in positions or movements where the robot vibrates very easily (extended configuration; short distances to travel) the antivibration control is more smooth and more accurate. The smoothness also means less resulting accelerations at the tool which leads to less mechanical stress.

The accuracy of circle paths however is not too good. This could lead to problems in applications like laser cutting or glueing where a very accurate path is needed with relatively high velocities.

5.2. Cycle Time

For this criterion the results for stabilization time and pose overshoot (see 4.2.1), the cycle time of the RIA test path (4.2.2) and the minimum positioning time (4.3.1) serve for a classification.

The stabilization time under torque control is *75 %* shorter and the pose overshoot is *87 %* smaller than under velocity control (RC3 - RC2).

The cyle time and traverse speed as well as the minimum positioning time are almost the same for RC2 and RC3. This occurs because velocity and acceleration are calculated in the same way in both controllers and also the time when an output is set depends on these calculations.

The results show that the robot moves with less vibrations under antivibration control. This could lead to faster cycle times. A welding gun for instance could be closed earlier because there has to be no waiting time until the vibrations stop.

Moreover, less vibrations causes less mechanical stress. This fact could be used in the design of new robot generations; smaller gearings, lightweight links etc.

Conclusion and Outlook

The SACODY project has succeeded in demonstrating the effectiveness of new robot control design methodologies and concepts as well as its relevance for upgrading the performance of industrial robot systems. The outcomes of the project are :

A set of software for the design of antivibration control

- The commercial LMS CADA-X software package for the Computer Aided Dynamic Analysis already integrates the Frequency Domain Direct Parameter Identification technique developed within the project.
- The BERTIN ADAMEUS software package for the dynamic simulation of multibody flexible mechanisms has been upgraded by new functionalities enabling model-based control engineering and verification by closed-loop simulations. These functionalities are intended to be integrated in the future software version to be released in 1993.
- The antivibration servo control software is now applicable to any industrial robot controller. Some preliminary estimates already available indicate that such a control applied in Automotive industry production units should enable 20 % time savings for spot welding cycle times.

A High Performance Servo Level environment integrated in a commmercial Robot Control

The successful results obtained with the industrial Servo Level for Robot Control developed by AEG enable to target its application to other flexible mechanisms such as large manipulators with electrohydraulic drives.

Sensor systems for Robot Evaluation

The K.U.Leuven have developed two prototypes of sensor systems one of which, the RODYM system, is now available on market.

Beyond these achievements, further exploitations of the project results are now being studied by the project partners :

The antivibration control technology will be integrated in the future controllers for the KUKA robots, with a view to upgrade their dynamic and operational performance.

The AEG servo level environment is already integrated in the AEG commercial robot controllers and machine tools. BERTIN will develop the exploitation of the ADAMEUS software now distributed by the FRAMASOFT company and the diffusion of dedicated antivibration control software for cranes, robot, telescopic structures, etc...

through bilateral contracts with SMEs and large companies.

K.U.Leuven will continue its effort for spreading the know-how developed in the framework of the project through their participation to seminar and conferences but also through their module in the Integrated European Course in Mechatronics.

Springer-Verlag and the Environment

We at Springer-Verlag firmly believe that an international science publisher has a special obligation to the environment, and our corporate policies consistently reflect this conviction.

We also expect our business partners – paper mills, printers, packaging manufacturers, etc. – to commit themselves to using environmentally friendly materials and production processes.

The paper in this book is made from low- or no-chlorine pulp and is acid free, in conformance with international standards for paper permanency.

GPSR Compliance

The European Union's (EU) General Product Safety Regulation (GPSR) is a set of rules that requires consumer products to be safe and our obligations to ensure this.

If you have any concerns about our products, you can contact us on ProductSafety@springernature.com

In case Publisher is established outside the EU, the EU authorized representative is:

Springer Nature Customer Service Center GmbH
Europaplatz 3
69115 Heidelberg, Germany

Batch number: 10704713

Printed by Printforce, the Netherlands